IMAGES
of America

Bingham Canyon Mine

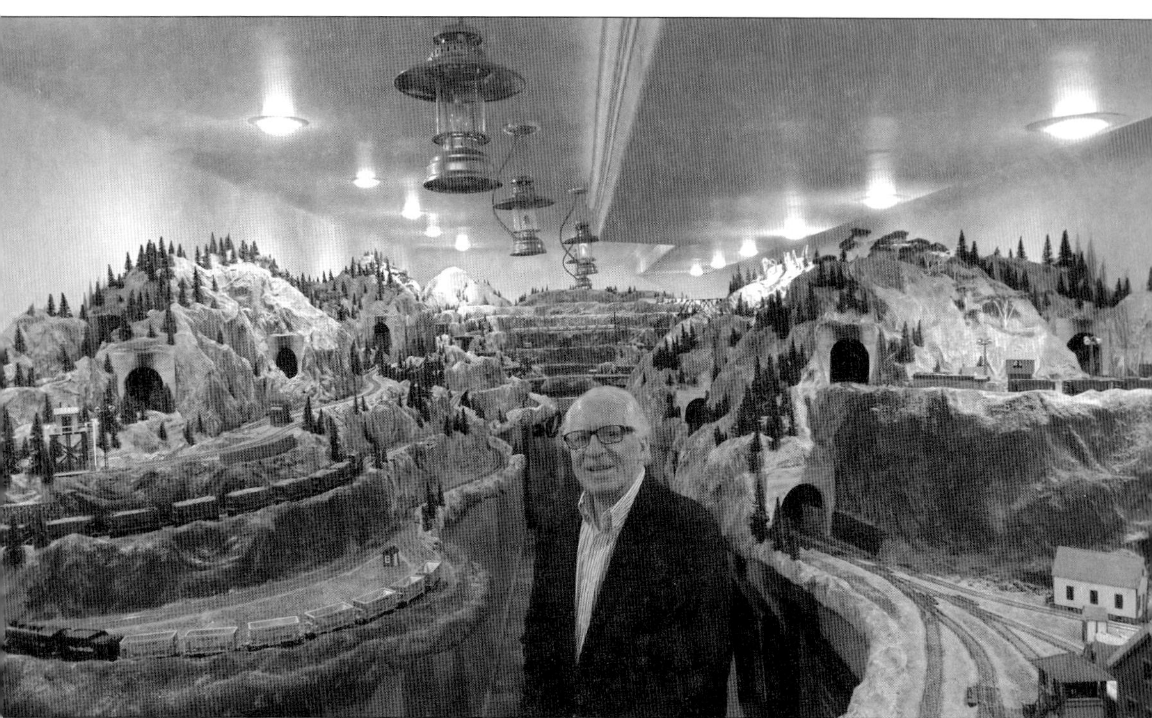

The author, so inspired by the workings of the Bingham Canyon Mine, built his own version of the Bingham Canyon Mine in his basement. His HO scale model railroad captures the mine when the railroad ruled the day, with trains and shovels constantly digging away at the many terraced levels. Looking at the mine with its size and vastness, it seemed to resemble a model railroad. (Author's collection.)

On the Cover: In this 1942 photograph, the brakeman is giving a hand signal to the engineer; the hand signal looks like the "highball sign" (meaning "All ahead is clear; let's go"). When the trains traveled cars first, the brakeman was required to ride on the front of the train. The background shows the many terraced levels of the awe-inspiring Bingham Canyon Mine. This massive pit was once a mountain; over the decades, it was eaten away by the ever-changing mining techniques. (Courtesy of Don Strack.)

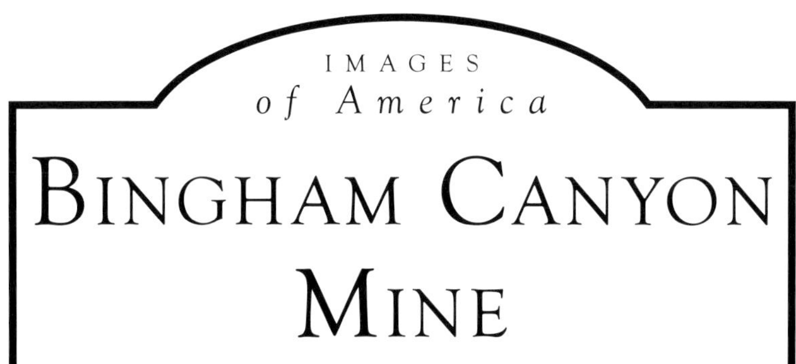

Tim Dumas

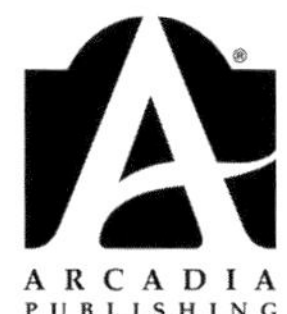

ISBN 978-1-4671-6304-0

Published by Arcadia Publishing
Charleston, South Carolina

Printed in the United States of America

Library of Congress Control Number: 2025941605

For all general information, please contact Arcadia Publishing:
Telephone 843-853-2070
Fax 843-853-0044
E-mail sales@arcadiapublishing.com

Visit us on the Internet at www.arcadiapublishing.com

To Kristine, my sweet wife, for her patience, support, and understanding of my passion for trains and Bingham Canyon, the mine, and its people

Contents

ACKNOWLEDGMENTS

A special thanks to Don Strack, as many of the images found in this book and much of the information came from him and his website (utahrails.net). This book is written in memory of Larry Sax (1937–2022); his collection of Bingham photographs and information is a priceless part of Bingham Canyon's history. Larry spent most of his life collecting all he could find out about Bingham and was always willing to share his collection and knowledge. I spent many hours scanning his information and listening to his many stories about Bingham and work at the mine. Larry and I thought that this information about Bingham was so important that we contacted Don Strack and asked if we could put this material on his website, where one can find 743 images of the Larry Sax collection. Many others saved pictures and documents when the mine was changing management and discarding this wealth of material. Thank you, James Belmont, Utah railroad photographer, for your wonderful photographs of the Bingham Canyon railroads and for being in the right place at the right time, capturing these important moments in Bingham Canyon's mine history. Thanks to the Bingham Canyon History Facebook group for all their wonderful knowledge of Bingham and the Bingham Canyon Mine and for sharing their photographs and stories—you are keeping Bingham Canyon history alive.

All images are from the Don Strack collection unless otherwise noted.

INTRODUCTION

The Bingham Canyon Mine is an awe-inspiring sight and has been from the beginning. Observation visitor centers have been set up and moved over time so people can view the mine. The mines' vastness and size cause a hypnotic stare as people look in amazement at this vast undertaking.

The story of the Bingham Canyon Mine is about constant movement and change, innovations, and improvements. Mining techniques and equipment were always improving.

Bingham Canyon Copper Mine is the largest man-made excavation on earth. There are larger mines covering miles of territory but saying it is the largest excavation means more material has been removed from this place than from any other place in the world. The massive work at the Panama Canal diggings, with the amount of earth moved, was overshadowed by the Bingham Canyon Mine by 1935. The mine was referred to as "the Hill" because there was once a mountain that was removed and then replaced with an enormous pit, making Bingham Canyon Mine one of the few man-made structures that can be seen from space.

Ore was discovered at Bingham Canyon in 1863, but before then, the Bingham brothers, Thomas and Stanford, settled in the canyon in 1848. The Bingham brothers only stayed in the canyon for a brief time, but Bingham would become the canyon's namesake. There were 43 years of underground mining before open-cut mining started at Bingham. By 1900, over 1,000 claims had been filed at Bingham; many mines changed hands and consolidated, but only a few were successful. At the turn of the century, the discovery of electricity and all its inventions created a great demand for copper, and Bingham had a mountain of it. A few men saw the potential of all this low-grade copper. It was Daniel C. Jackling's idea to mass-produce low-grade copper by using steam shovels and trains. Jackling was instrumental in getting investors. He ran tests with an underground mine and a test mill (the Copperton Mill), and with the results, his vision was realized.

Open-cut mining started in 1906 at Bingham, and both Daniel Jackling's Utah Copper and Samuel Newhouse's Boston Consolidated Mining Company started digging away on the same mountain for copper. The companies merged in 1910, continuing to operate under the Utah Copper name. It was the first time this mining method was used with copper. Critics said this kind of high-mountain, open-cut mining could not be done; this was hard rock mining, every inch had to first be blasted, trains and shovels had to tackle the mountain, and then ore had to get down the mountain to the main line and to the mills. Jackling and his engineers were always looking for a more efficient and economical way to move ore, and there were always problems to solve. These men wrote the book on low-grade copper mining. Changes both big and small were always happening at the mine. This book is about some of those changes.

One of the first big changes that came to the Bingham Canyon Mine was the building of its own railroad, the Bingham & Garfield Railway (B&G). At the startup of open-cut mining, both Utah Copper and Boston Consolidated Mining Company contracted with Rio Grande Western Railway (RGW) to move the low-grade ore from the mine to the mills, hence the name "the low-grade line." From the beginning, in 1906, the Rio Grande Western Railway could not keep up with the volume of material coming out of the two mines. It refused to make improvements, and without a good response from Rio Grande Western, Utah Copper organized the Bingham & Garfield Railway in 1908, with construction starting in 1910. With Utah Copper owning its own railroad, it would prove to be successful. Another early advancement was changing the first steam shovels; the early steam shovels operated on parallel standard-gauge tracks alongside the trains. The shovels were fitted with crawler treads, which made the shovels more versatile. Advancements in electric motor technology enabled the conversion from steam to electricity, first with shovels and then with trains. Both DC and AC currents were tried on the electric shovels; eventually, AC current won out. The trains ran on 750-volt DC power. The electric railroad would become the largest privately owned electric railroad in the world. As the mine expanded more, dumping grounds were needed, so mine management was always buying up old claims with surface rights for a place to dump its overburden. To access dumping grounds on the north side of Bingham Canyon, Utah Copper had to build large bridges that crossed over Carr Fork Canyon and tunnels for more dumping grounds as it moved down Bingham Canyon. Once the pit started to be formed and the trains had to pull the ore up out of the mine, three railroad tunnels were eventually built in the mine to keep the ore moving downhill. Advancements in large trucks and tires led to the top two-thirds of the mine becoming shovel-to-truck operations. After 40 years of great service, the electric railroad was converted to diesel power.

In the olden days, when a mine failed, the town around it became a ghost town; however, it was the opposite with the Bingham Canyon Mine, as the surrounding towns became ghost towns with the success of the mine. The first town to go was Copperfield, which sat on the southeast corner of the mine. The next community to be gobbled up by the mine's expansion was Highland Boy. Highland Boy was on the north side of the mine at the top of Carr Fork Canyon. The entire Carr Fork Canyon, including the community of Highland Boy, will be dug away. The last and largest town devoured by the mine's insatiable need for copper was Bingham. In 1971, Bingham was disincorporated, with only a few people and buildings remaining. Eventually, everything in Bingham Canyon was removed, eaten away, or buried.

When people look west at the Oquirrh Mountains and see the large mining operation, they see a scarred mountain and an environmental eyesore, but there is a lot more to the story than meets the eye. It is our voracious demand for copper, more cars, bigger houses, and more electronic devices that has destroyed the mountain and buried the canyon. Copper has changed our lives for the better. Copper is the best affordable conductor for electricity; it is pliable, easy to wrap and solder, and the safest and most efficient metal. When I look at Bingham Canyon Mine, I see history, a better wage, and improved living; the people that once lived and worked at Bingham and the many immigrants (at one time 65 percent of Bingham's people were foreign born) living and working together in harmony; and people that were happy and content with less but worked hard to better themselves and their families. Bingham Canyon Mine was once the largest employer in the state and the largest taxpayer. The wages changed people's lives for the better. Many people worked for Kennecott—the mine, mill, or smelter—and if one did not work here, they probably knew someone who did. The resulting wealth of the Bingham Canyon Mine has earned it the reputation as "the richest hole on earth."

There is a huge mark on the mountainside, but it should be a reminder of what it has given and not just what it has taken: the Bingham Canyon Mine.

One

Discovery and Underground Mines

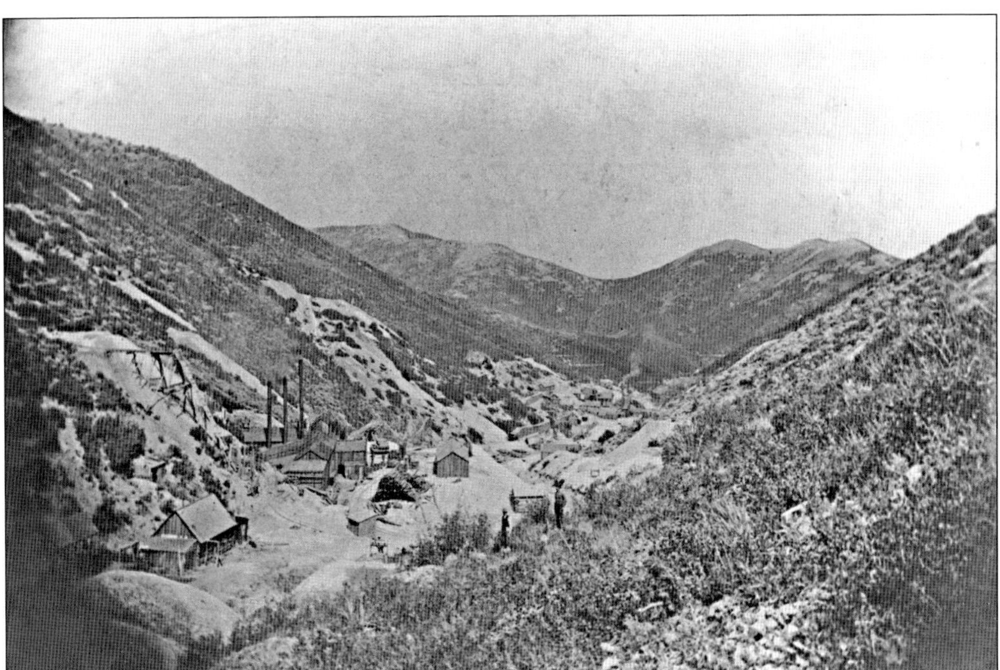

This image shows Galena Gulch of Bingham Canyon. This was the site of the first mining claim, not only at Bingham, but also in the territory of Utah in 1863. The canyon's name was derived from the Bingham brothers, who settled the canyon in 1848 after being sent there by Brigham Young, the prophet of the Church of Jesus Christ of Latter-day Saints, who led the saints to the Salt Lake Valley in 1847.

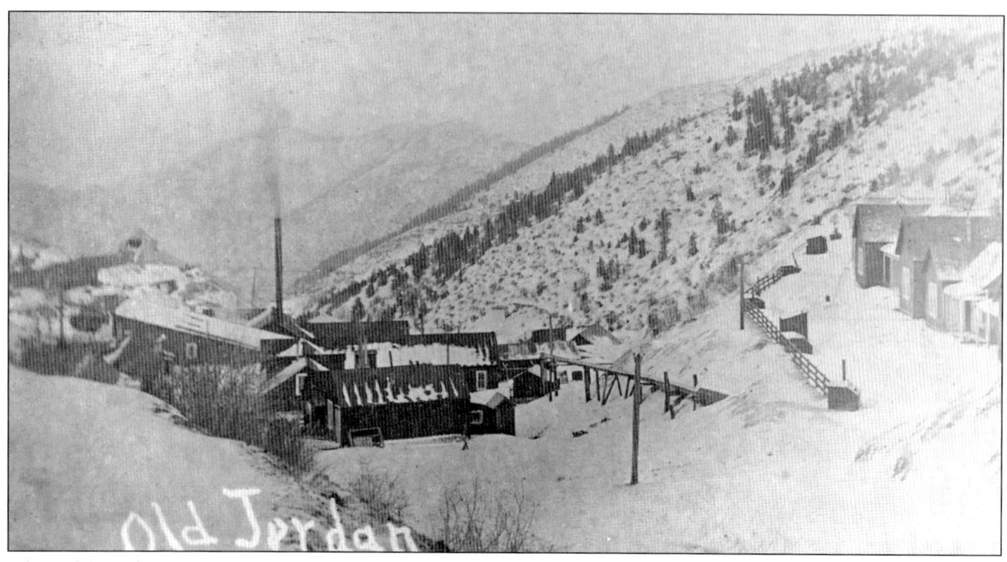

The Old Jordan Mine was first staked on September 17, 1863. One of the more credible versions of the discovery is that George B. Ogilvie, a logger working for Archibald Gardner, noticed shiny rocks that were dug up by logs being pulled down to Gardner's sawmill. Ogilvie was advised to take the rocks to Fort Douglas commander Col. Patrick O'Connor, who confirmed that the rocks were ore.

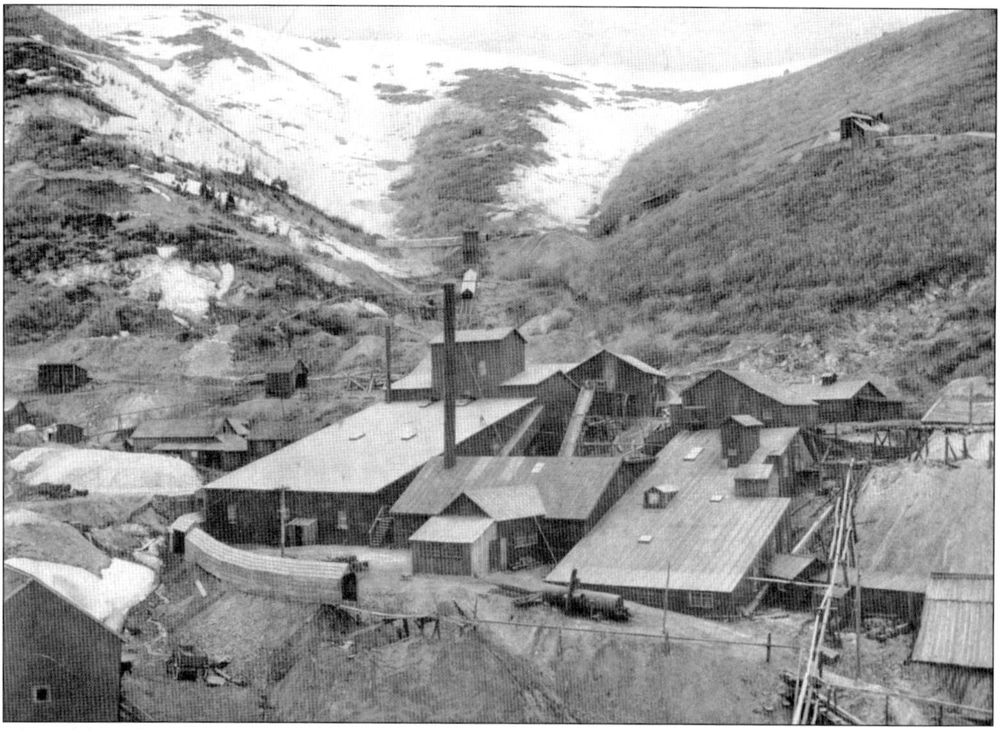

The Old Jordan Mine is seen here in 1900. On September 17, 1863, twenty-five men—Colonel O'Connor and soldiers and Archibald Gardner and loggers—met in West Jordan, recorded the first claim, and organized the Jordan Silver Mining Company. They also established the West Mountain Mining District; this district encompassed the entire Oquirrh Mountain range from the Great Salt Lake on the north to the Utah Lake on the south.

This image shows Galena Gulch in 1891 with the Old Jordan and Galena Mines; these were two separate mining claims. By late 1873, they were being worked as one single mine. After the discovery of ore in Bingham in 1863, a flood of prospectors came to Utah Territory seeking their fortune. They explored the mountains around the Salt Lake Valley. In the Wasatch Mountains, more ore was found at Park City and Alta.

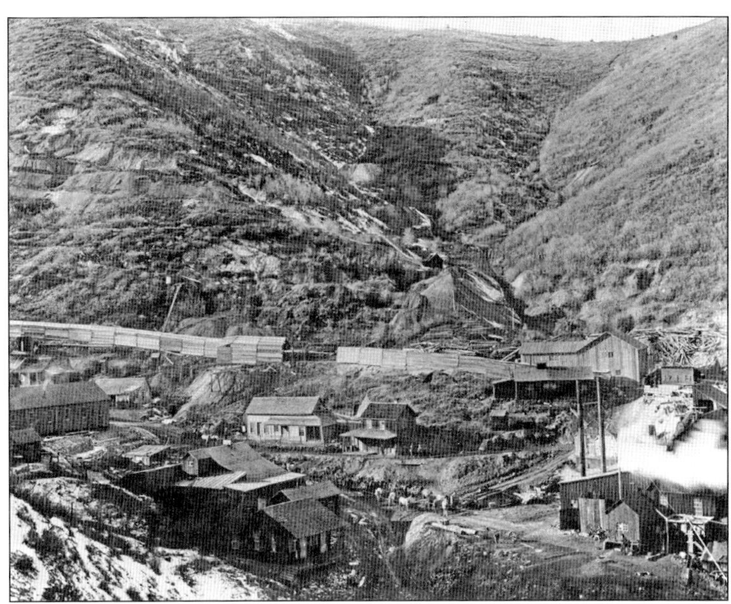

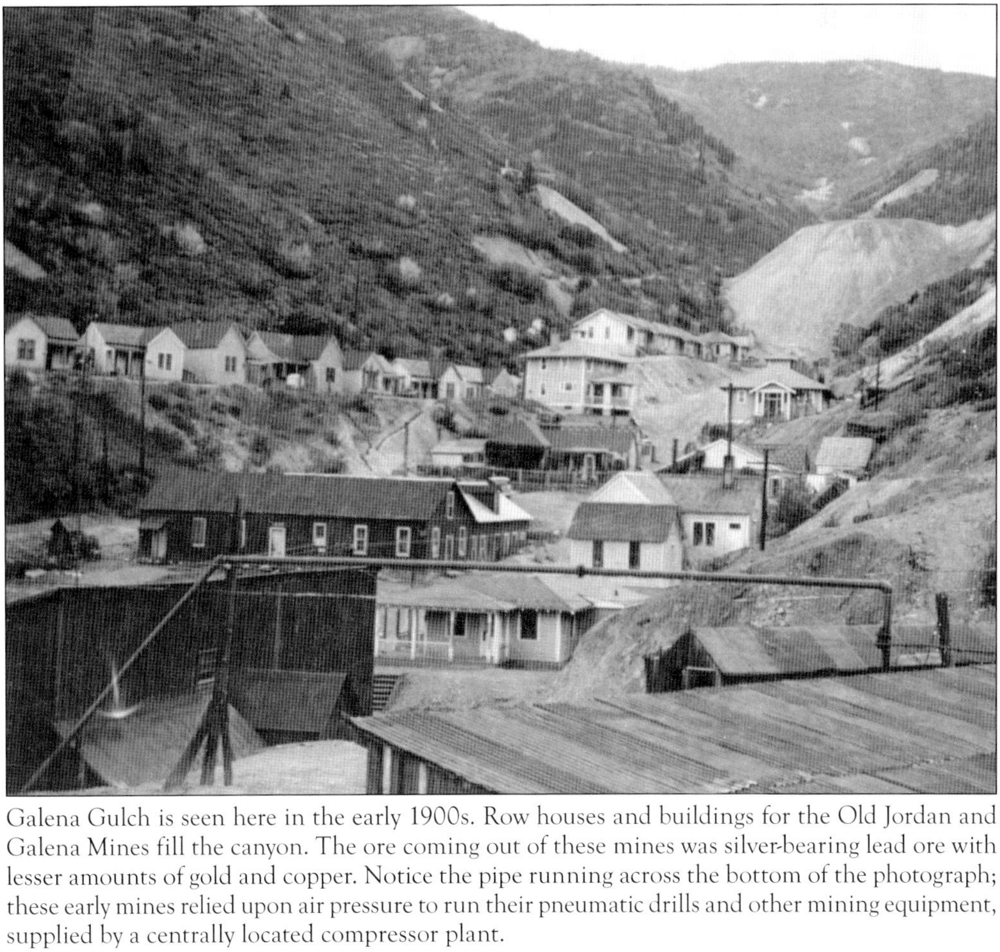

Galena Gulch is seen here in the early 1900s. Row houses and buildings for the Old Jordan and Galena Mines fill the canyon. The ore coming out of these mines was silver-bearing lead ore with lesser amounts of gold and copper. Notice the pipe running across the bottom of the photograph; these early mines relied upon air pressure to run their pneumatic drills and other mining equipment, supplied by a centrally located compressor plant.

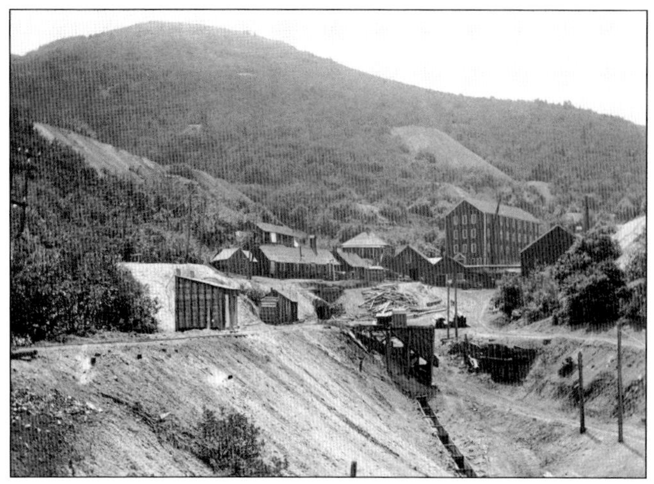

The Old Jordan and Galena Mines are pictured here in 1914. As the early mines at Bingham developed and grew, they started to overlap into each other's claims. One solution was to consolidate properties. In March 1899, the US Mining Company was organized, combining not only Old Jordan and Galena Mines but Northern Light, Orphan Boy, Live Pine, Utah, Niagara, Spanish, and Old Telegraph Mines, to name a few. Albert F. Holden was the managing director.

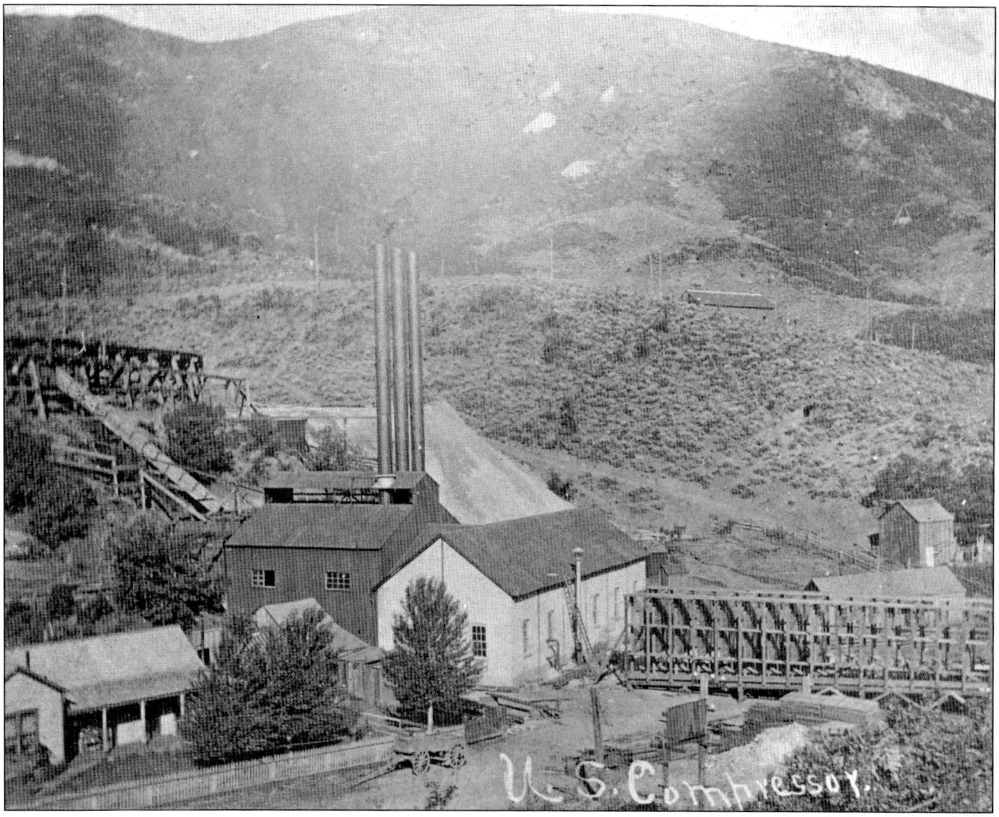

The US Mining Company compressor plant in Upper Bingham is seen here. Upper Bingham was renamed Copperfield in 1914, and the main canyon branched off into places like Copper Center Gulch, Galena Gulch, Bear Gulch, and Copper Gulch. From the compressor plant, a series of pipes went up the mountainside to the different US Mining Company mines. The mountain in the background is where open-cut mining would take place, starting in 1906.

This image, taken in 1942, shows the US Mining Company Niagara tunnel, center right, in the town of Copperfield. By 1914, with the consolidation of the US Mining Company mines, the higher mines in Galena and Bear Gulches were connected with a system of tunnels; they used the Niagara tunnel as the main transportation tunnel to carry the ore out. The large E-line bridge provided access to the Denver & Rio Grande Western Railway (D&RGW) tracks to the Midvale smelter.

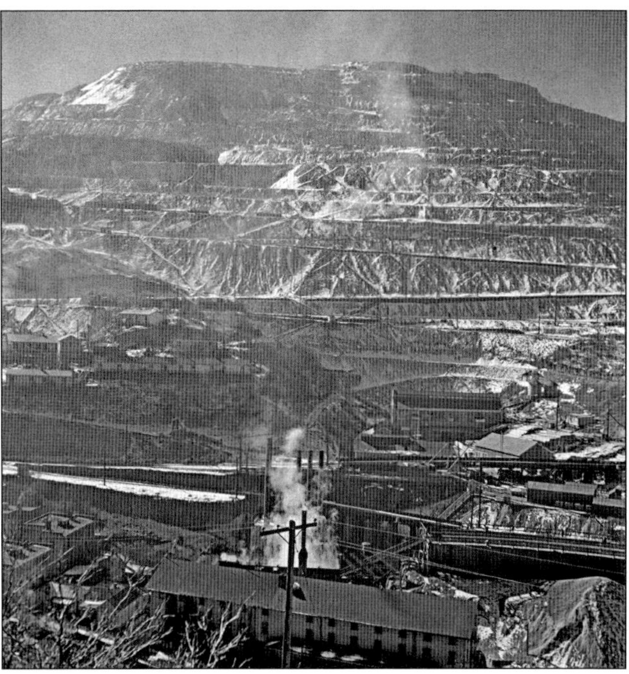

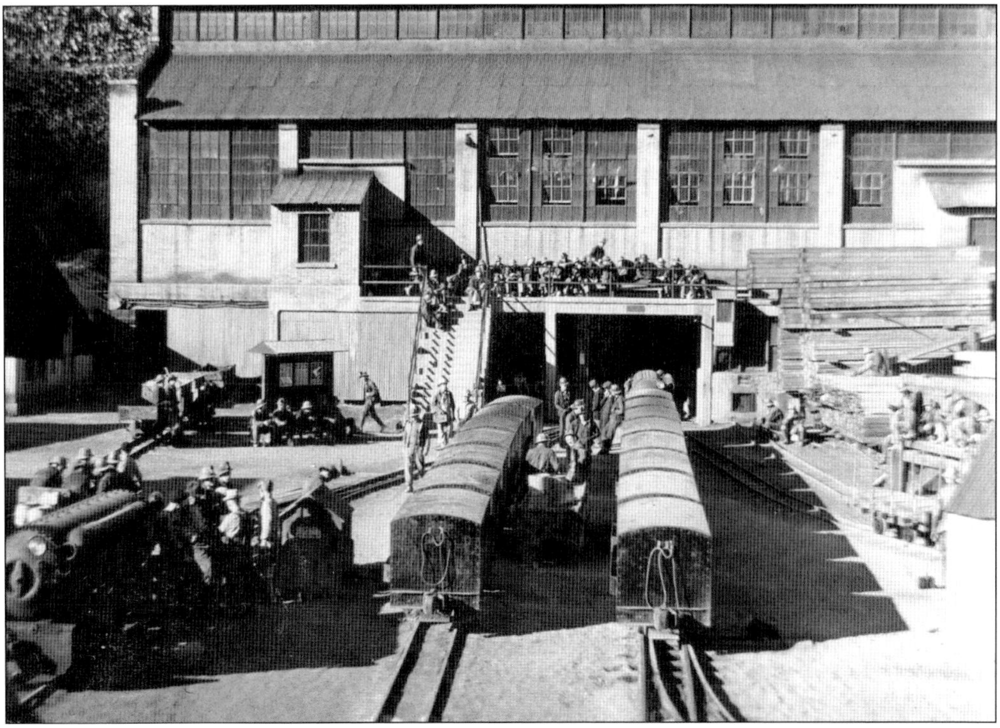

This is a close-up view of the Niagara tunnel during shift change. With the completion of the Bingham-Lark tunnel in 1952 at Lark, Utah, the US Mining Company operations in Copperfield would be no more. The new tunnel connected all the US Mining Company network of tunnels with the new opening at Lark. The Bingham-Lark tunnel was 21,300 feet in length and cost $6 million to build.

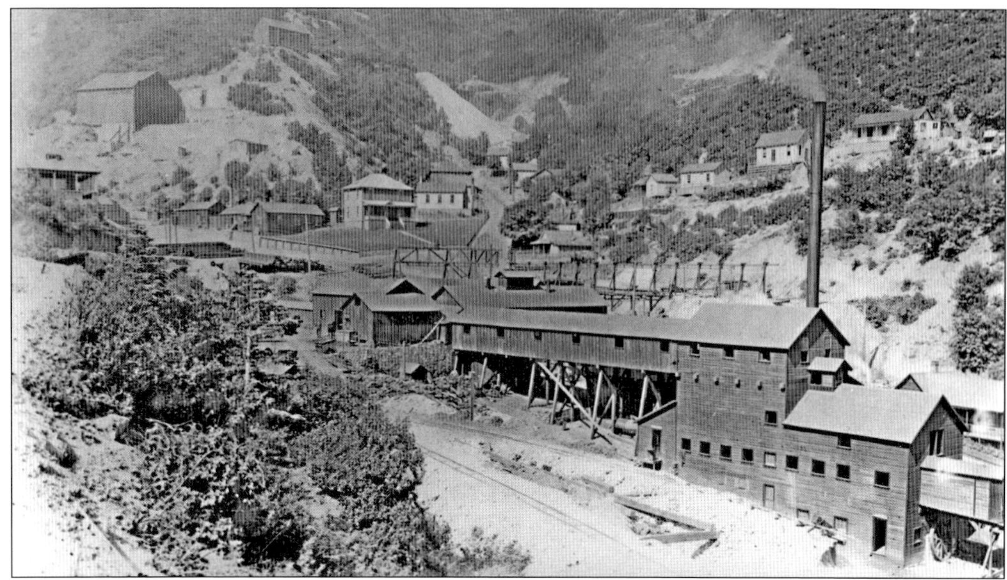

This image shows the Highland Boy Mine at the top of the Carr Fork Canyon. Note Sap Gulch above the mine, top center. James W. Campbell first staked the Highland Boy claim in 1873. Samuel Newhouse and associates purchased a group of 10 mining claims, which included the Highland Boy and Stewart Mines, in 1896 at the cost of $200,000. Thomas Weir was appointed manager of the properties, and Samuel Newhouse made improvements, including the addition of a gold cyanide concentration mill, seen top left.

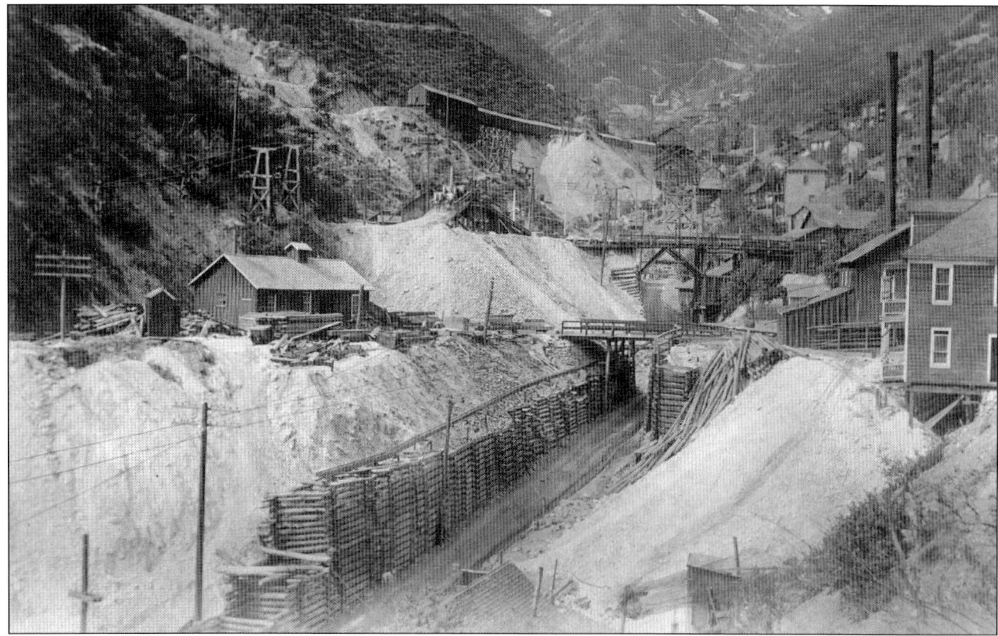

This view is looking up Carr Fork Canyon to the Highland Boy Mine properties. With Samuel Newhouse, the name changed to Utah Consolidated Gold Mines. On February 21, 1899, Standard Oil purchased a controlling interest for $6 million, and it became Utah Consolidated Mining Company. Finding extensive copper deposits, some as high as 25-percent copper, the Highland Boy Mine became the leading copper producer at Bingham.

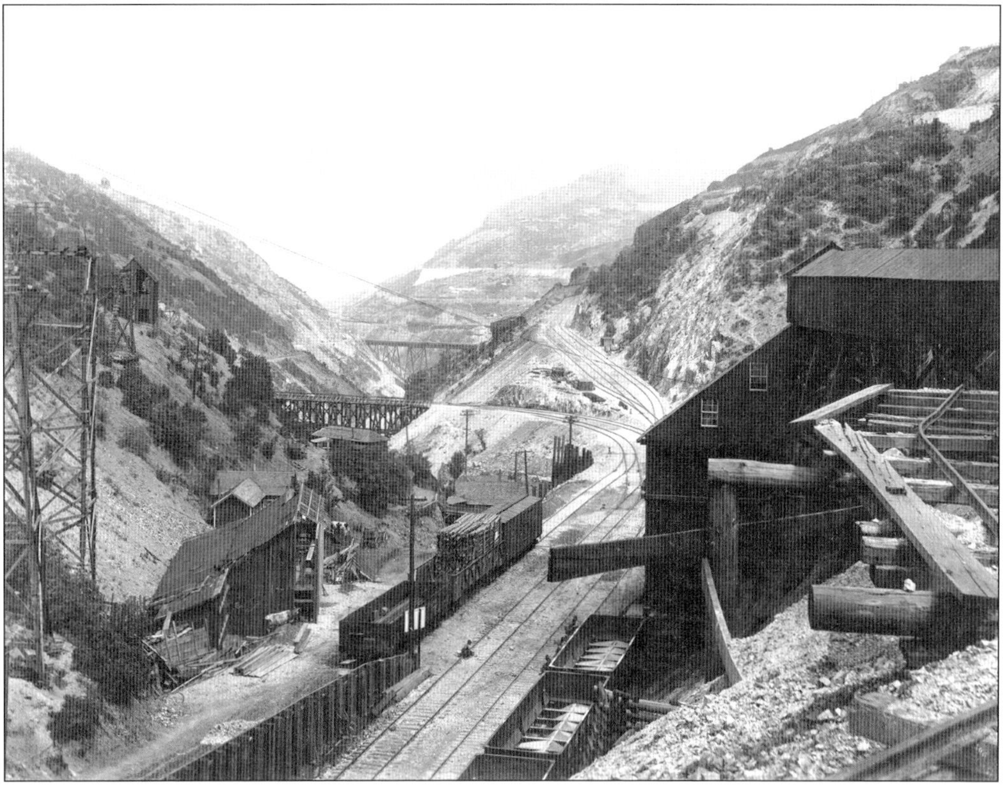

This image is looking down Carr Fork Canyon in 1914 at the Utah-Apex workings. Utah-Apex was incorporated in May 1902 with a merger of Copperfield Mines and York Mining Companies, which consisted of 33 claims on 254 acres. The claims were located on York Hill, with Carr Fork Canyon adjoining the Utah Consolidated claims. This area in Carr Fork Canyon was referred to as "Phoenix," home of the Phoenix tunnel and mill.

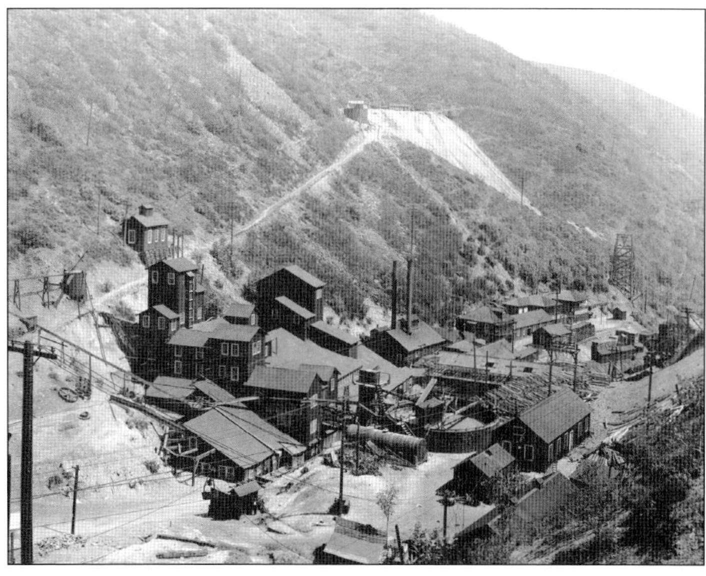

This view is looking down on the Utah-Apex complex. The court fights between Utah-Apex and Utah Consolidated were one of the largest in Utah mining's history, eventually breaking both companies. Both closed their mines through the 1920s and 1930s because of the court fights and low metal prices. In 1937, Utah-Apex merged with Utah Delaware (the successor to Utah Consolidated) to form National Tunnel and Mines Company.

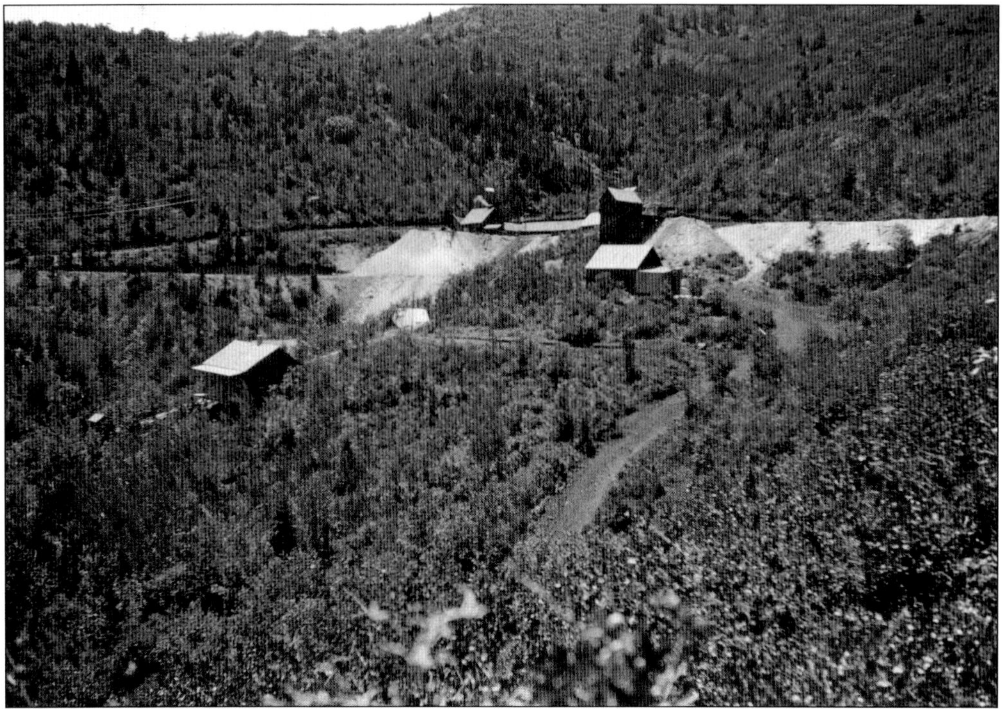

This image shows the Commercial Mine in 1903, located in Copper Center Gulch. It became part of the Bingham Mines Company in 1908, reorganized as Bingham Consolidated Mines and Smelter Company in April 1908. Utah Copper acquired surface rights on September 24, 1910, for dumping grounds in Copper Center Gulch. The Commercial Mine would move 12 buildings down half a mile to a place near the Niagara tunnel.

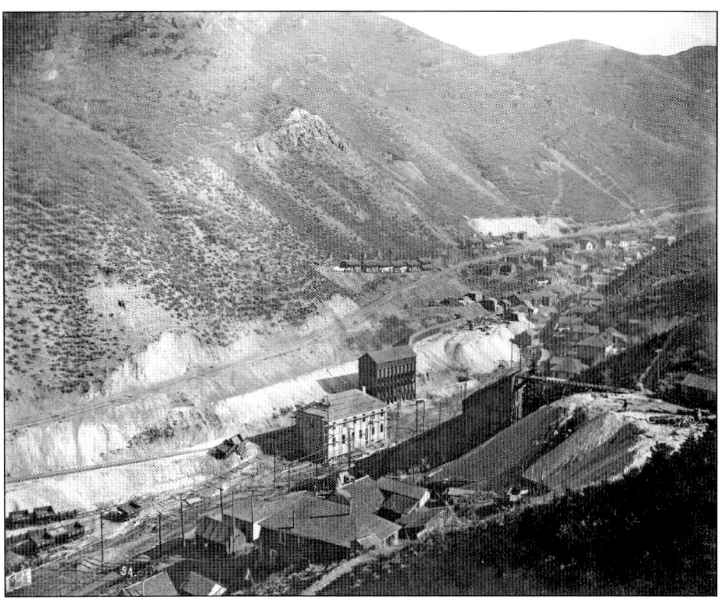

This photograph shows the workings of the Ohio Copper Company in 1906. The Ohio Copper Company was organized in October 1903 to work 120 acres from the Columbia and Erie claims. They were trying to mine the low-grade copper ore. Note the Ohio Copper's loading station and the six row houses in the center of the photograph above it. The row houses became a landmark in early Bingham Canyon.

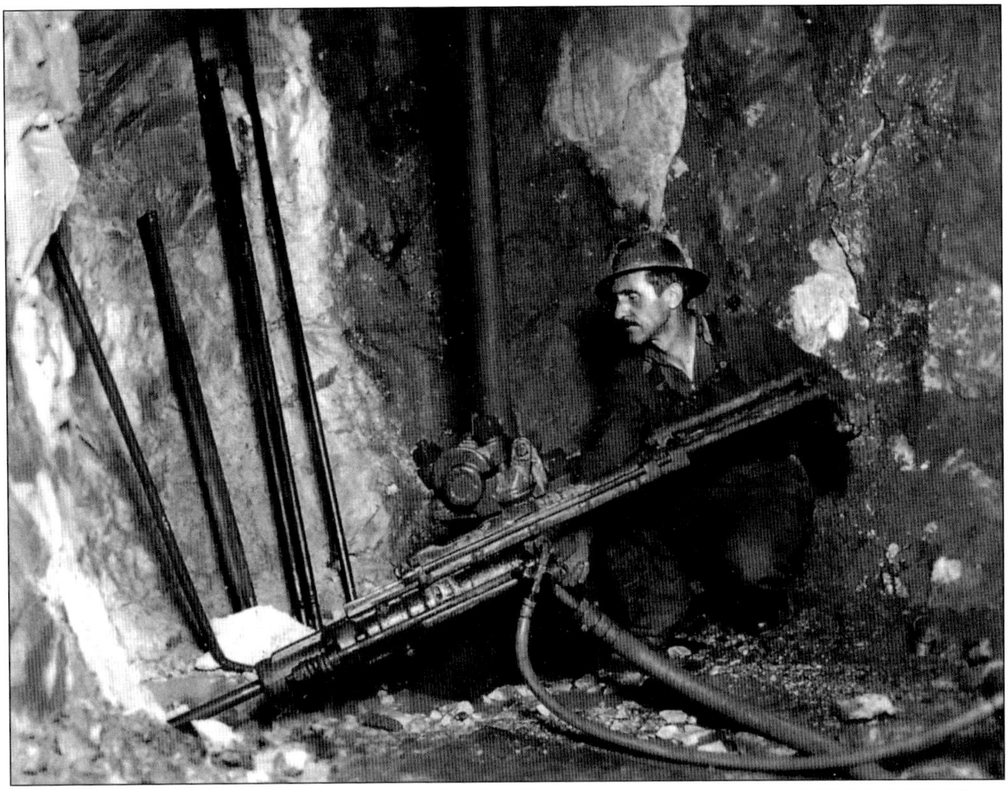

Seen here is an unidentified underground hard rock miner using a pneumatic drill (air drill). These drills created dust and, in this limited space, produced a dust cloud. The dust was fine particles that contained toxic heavy metals and other hazardous materials such as silica, lead, and arsenic. This work caused irritation to the eyes and nose as well as more serious upper respiratory problems such as asthma, bronchitis, and silicosis, a lung disease.

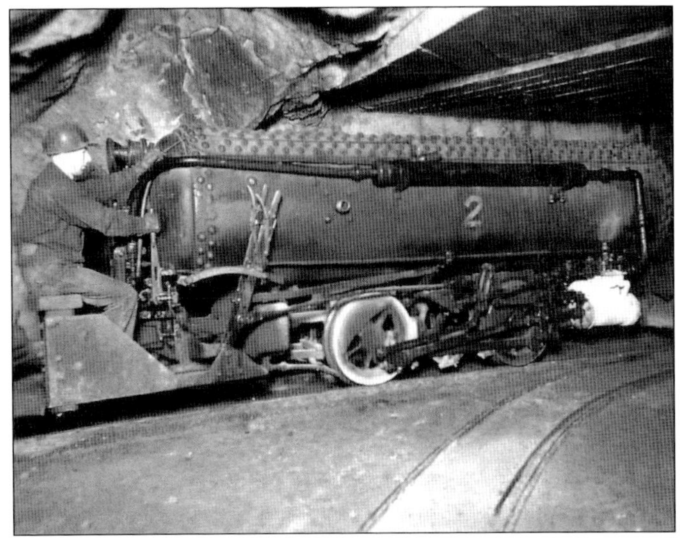

US Mining Company operations used eight-ton Porter compressed air locomotives in its underground mining. In 1914, the US Mining Company stopped using its aerial tram to transport ore to the D&RGW trains at Frog Town. Expanding the Niagara tunnel, it was now used to bring the ore out of the mine and then reload it into railcars. Using compressed air locomotives or electric locomotives helped keep dangerous fumes out of the mine.

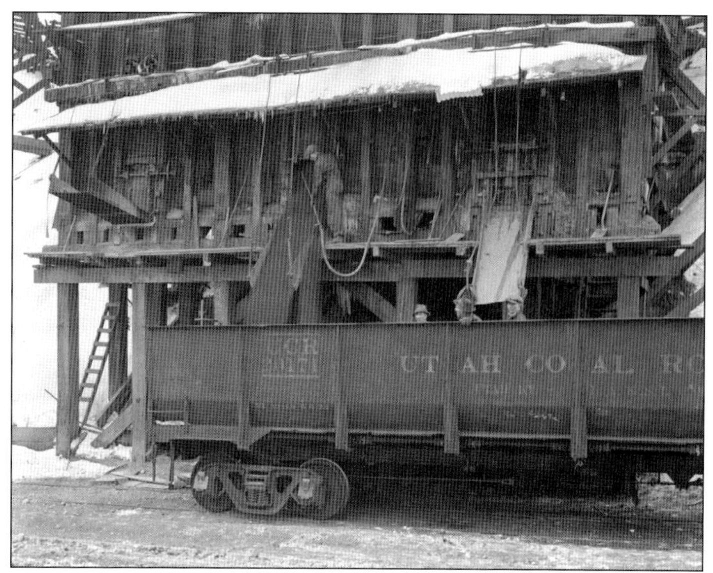

Seen here is the Utah-Apex loading station at Carr Fork Canyon. This scene played out many times around Bingham. With all the underground mines, many had a loading station where the ore was brought to be loaded into a train car. The image shows the railcar as being that of Utah Coal, but it was used by the Denver & Rio Grande Western Railway to transport ore. Note the men inside the railcar, possibly situated to level out the load.

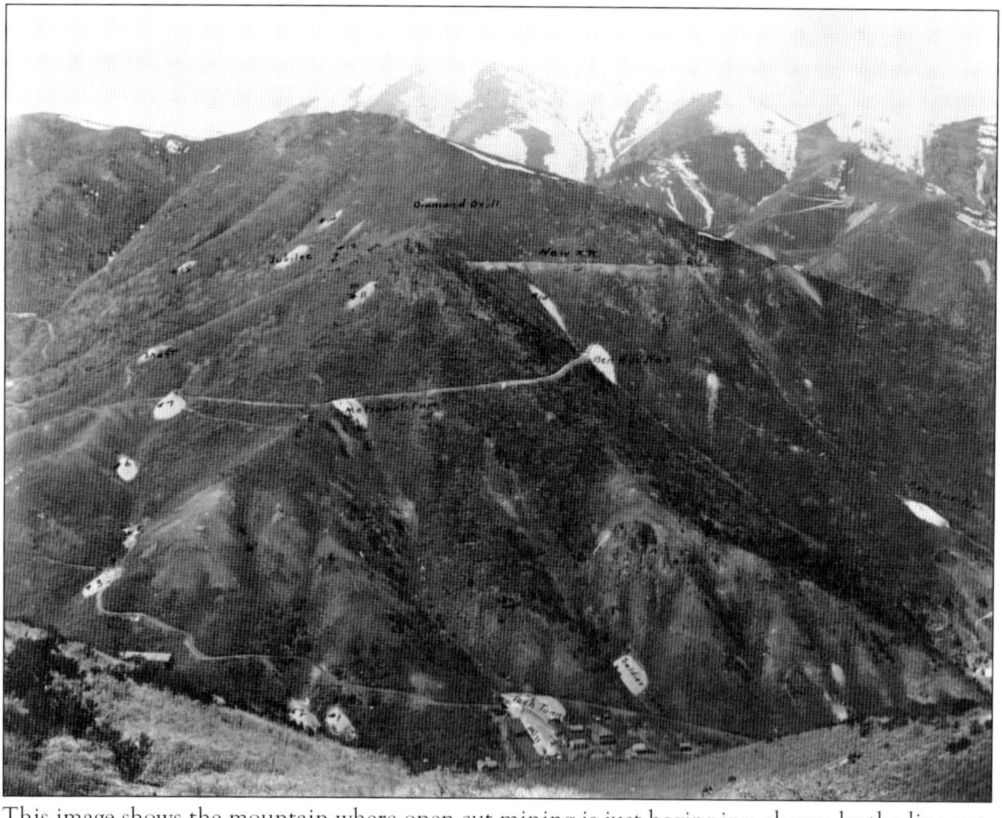

This image shows the mountain where open-cut mining is just beginning, shown by the line cut, top right (labeled "new line"), and workings, bottom right. Now, the mountainside is filled with underground mines, and their mine dumps show their location. The photograph is labeled with names of a few of these mines, such as Tech tunnel, Soldier tunnel, Metropolitan, Ben Hur Nos. 1 and 2, Jubilee, and Diamond Drill. Copper Center Gulch, home of the Commercial Mine, is left.

Two

Trains to the Canyon and Ore to the Trains

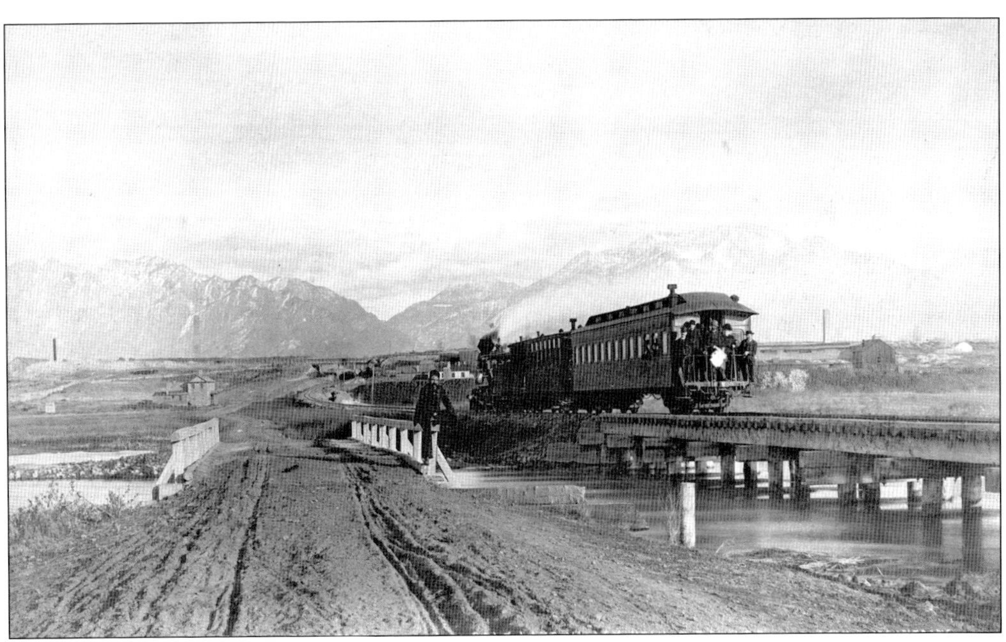

This image shows the Bingham Canon (old spelling) & Camp Floyd Railroad. The railroad was narrow gauge and completed on November 22, 1873. The passenger train is crossing over the Jordan River heading for Bingham Junction (Midvale). The Bingham Canyon & Camp Floyd Railroad is 16 miles from Sandy Station to Lower Bingham. The railroad was purchased by the Denver & Rio Grande Western Railway in 1881.

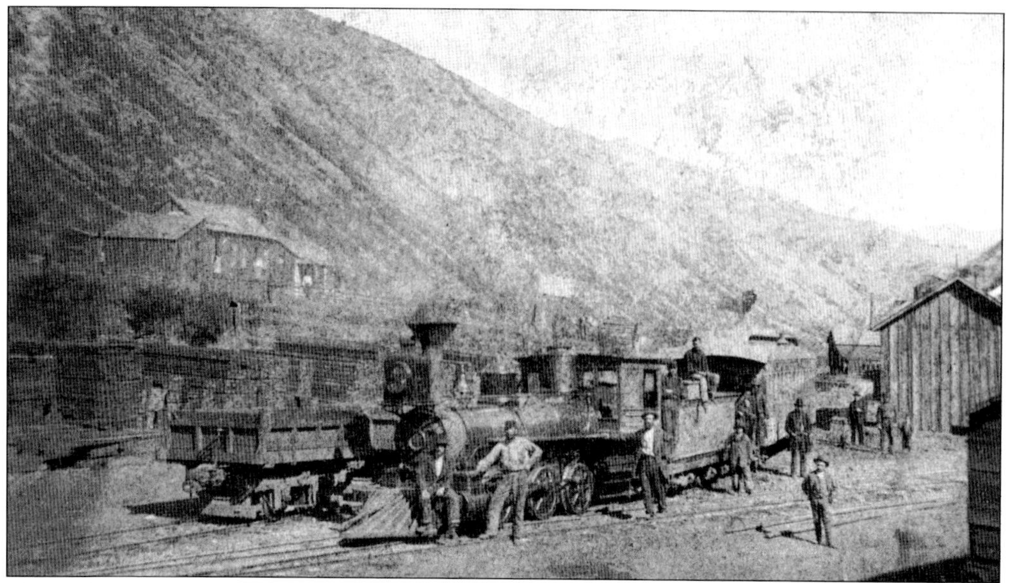

This old photograph shows a passenger train coming into Lower Bingham (Frog Town). The Bingham Canyon & Camp Floyd Railroad was narrow gauge, but after Denver & Rio Grande Western purchased the property, it upgraded the line to standard gauge. This image probably shows the first standard train entering Frog Town on June 3, 1890. The underground mines at Bingham were in a recession due to the cost of transportation until the railroad came.

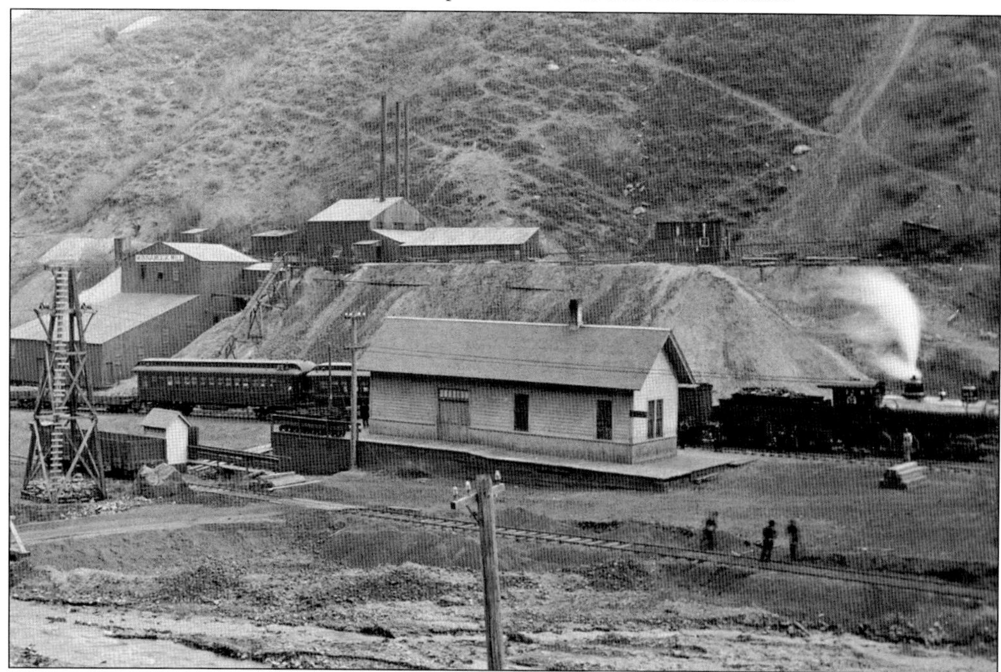

This image shows the new D&RGW depot at Frog Town, with the Winnamuck Mine and mill in the background. The old depot was two narrow-gauge boxcars that functioned as ticket and express offices. No one would sell the railroad a parcel of land for a new depot for a reasonable price. Finally, the Winnamuck Mine sold a 60-by-100-foot tract of land for the depot; the new depot was completed on June 25, 1896.

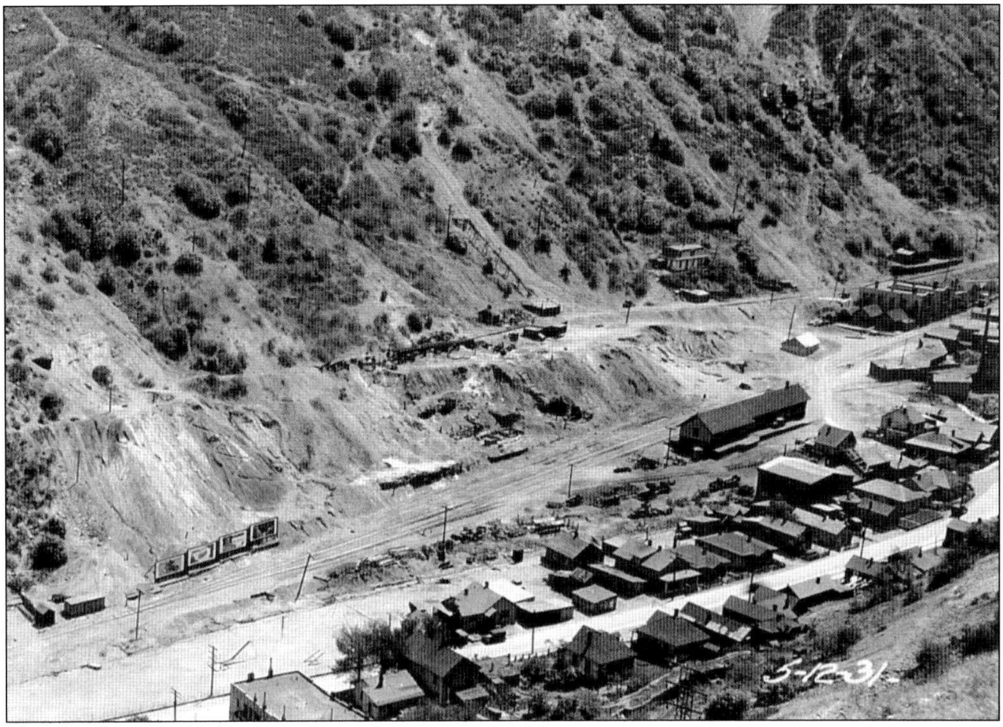

This image was taken at Frog Town, also named Lower Bingham, on May 12, 1931. The D&RGW train depot is pictured center right, and the Alexander Apartments are top right. Houses and boardinghouses line the canyon floor. The remains of the Winnamuck Mine and mill are seen in the center. The south side of the canyon is being cleared for the cross-canyon connection, and the new route to the mills that would be built in the early 1940s; it was part of keeping the ore moving downhill.

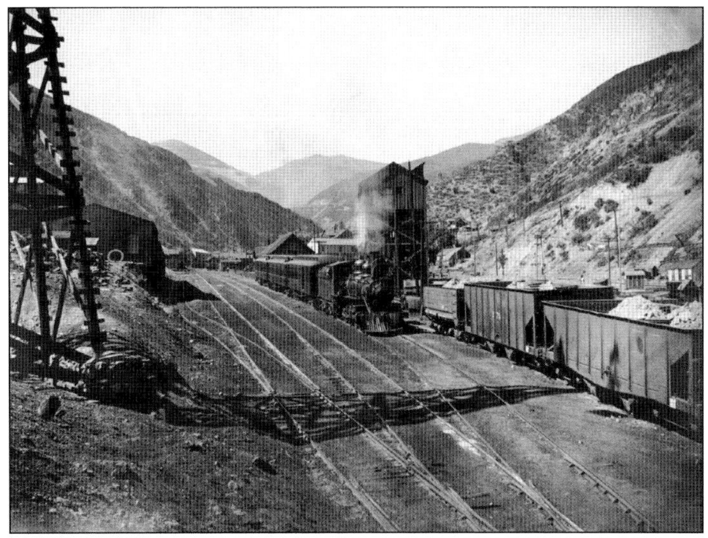

A passenger train leaves Frog Town. The Winnamuck Mine and mill are seen to the left. The large aerial tram tower, seen left, serviced the US Mining Company mines in Upper Bingham. The Highland Boy aerial tram building is in the center of the photograph. There were once four aerial trams in Bingham because the trains from the valley stopped at Frog Town, and ore from the mines had to work its way down to the trains.

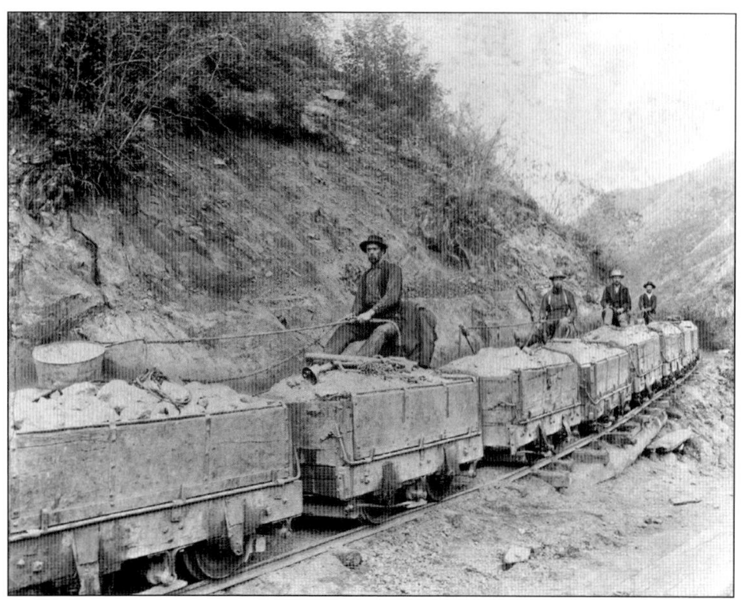

Pictured is a gravity rail tramway. Men riding on the ore cars controlled the downhill speed by using hand brakes for the trip down. Horse and mule teams pulled the empty cars back uphill to the mines. The gauge of the rails is in question; it was somewhere between 24 inches and three feet. The first ore coming out of the mines at Bingham used a horse and wagon, making it a slow and arduous journey.

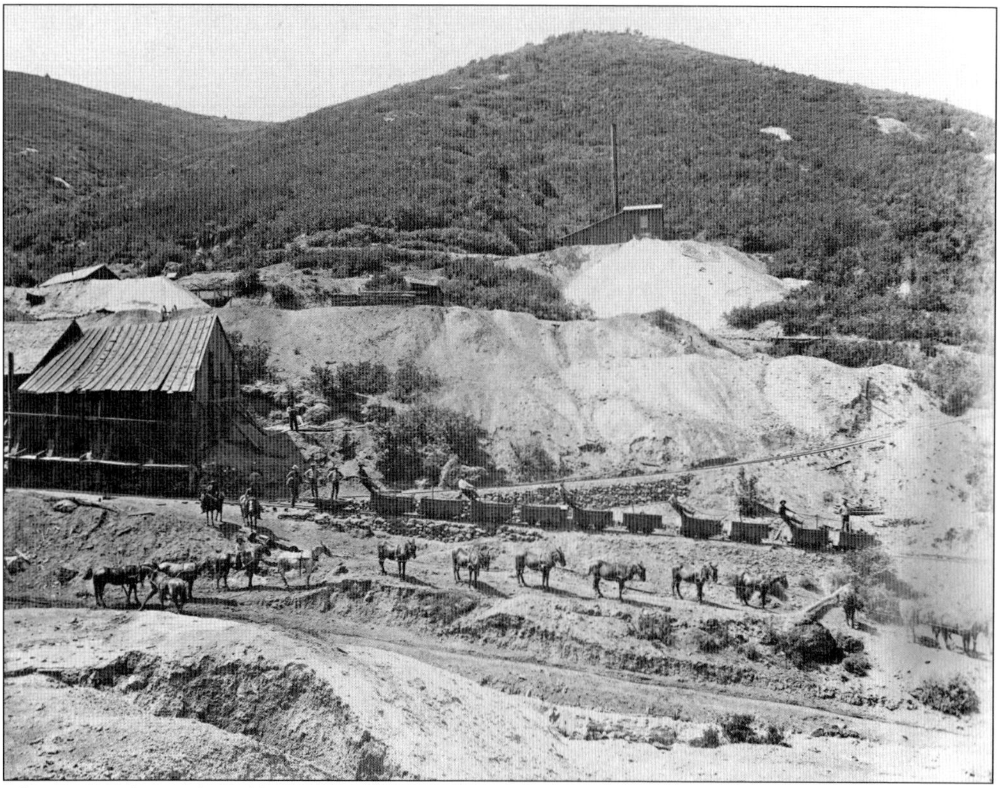

This image shows the Old Jordan Mine in 1889. Men pose on their ore cars for the trip down the canyon. The first gravity rail tramway was completed in 1875. It traveled from the Old Jordan and Galena Mines, down three miles to the station at Frog Town. Over time, the rail tram was extended to the Commercial Mine and others. The grade to the Commercial Mine was steep at 7.4 percent.

This photograph shows a gravity rail tramway running down the canyon in the center. Miners are dumping ore into the ore bin; note the mine opening down the way. The gravity rail tramway was a good system when there were fewer amounts of high-grade ore to be moved. Each of the rail tramway cars held three tons. The image shows 12 ore cars on the tramway; the cars would make two trips a day.

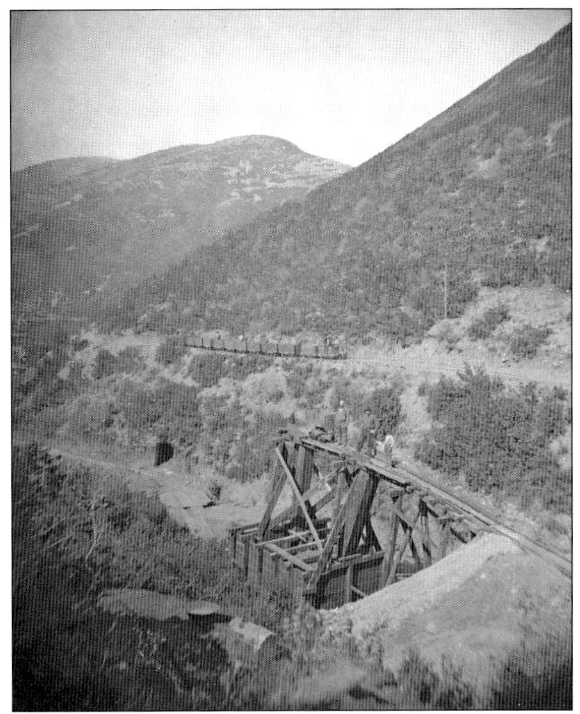

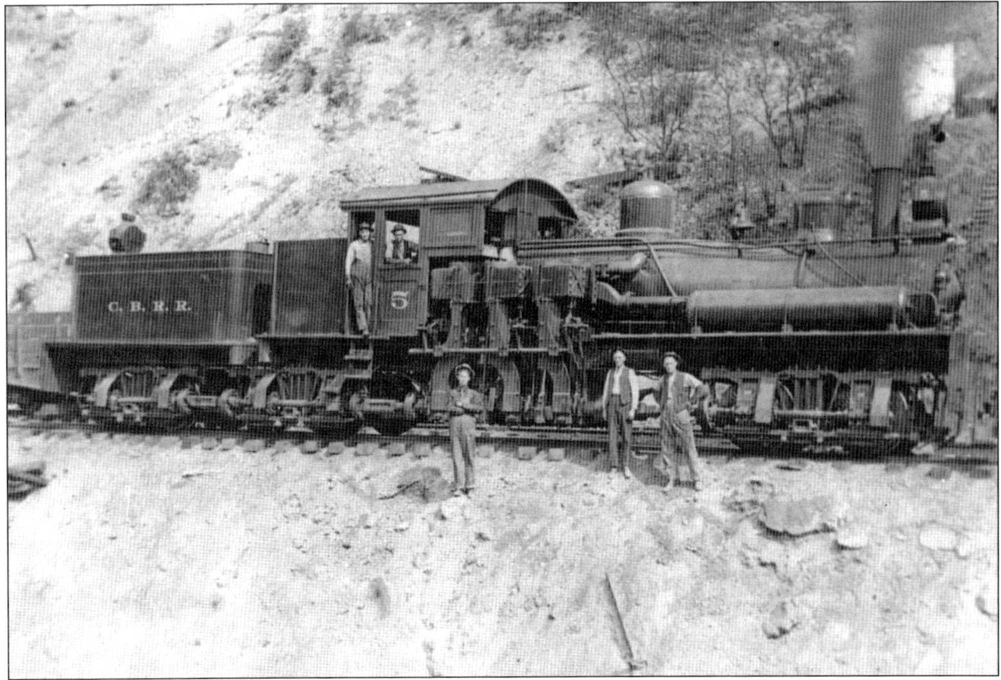

A Shay locomotive is shown on the Copper Belt Railroad. The gravity rail tramway was replaced with the standard-gauge Copper Belt Railroad. The old rail tramway route was widened and improved with help from the Rio Grande Western Railway. The new type of Shay locomotive was a good fit for the steep grades and sharp turns running up to the higher mines in Bingham Canyon. Ephraim Shay was the inventor, and it was patented in 1881.

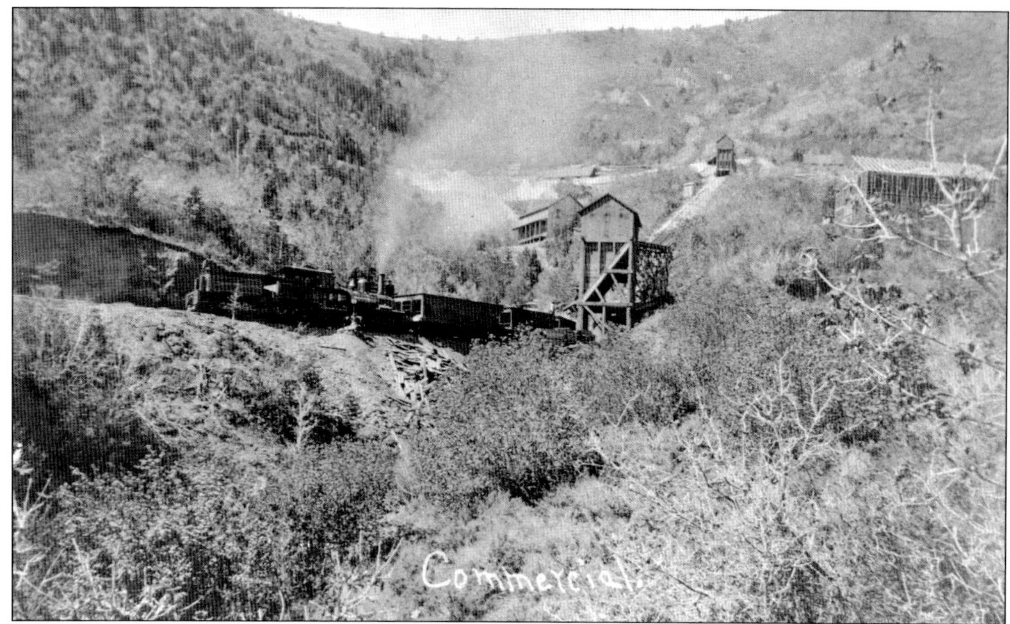

Shay locomotive No. 1 is pictured at the Commercial Mine in Copper Center Gulch. Lima Locomotive Works built the locomotive in April 1900; it was delivered April 26, 1900, to J.G. Jacobs for his Salt Lake & Mercur Railroad. It was transferred to the Copper Belt Railroad at Bingham on January 5, 1901. By 1906, there were five Shay locomotives at Bingham. All five were built by Lima and were standard Class C, three-cylinder, three-truck Shay locomotives.

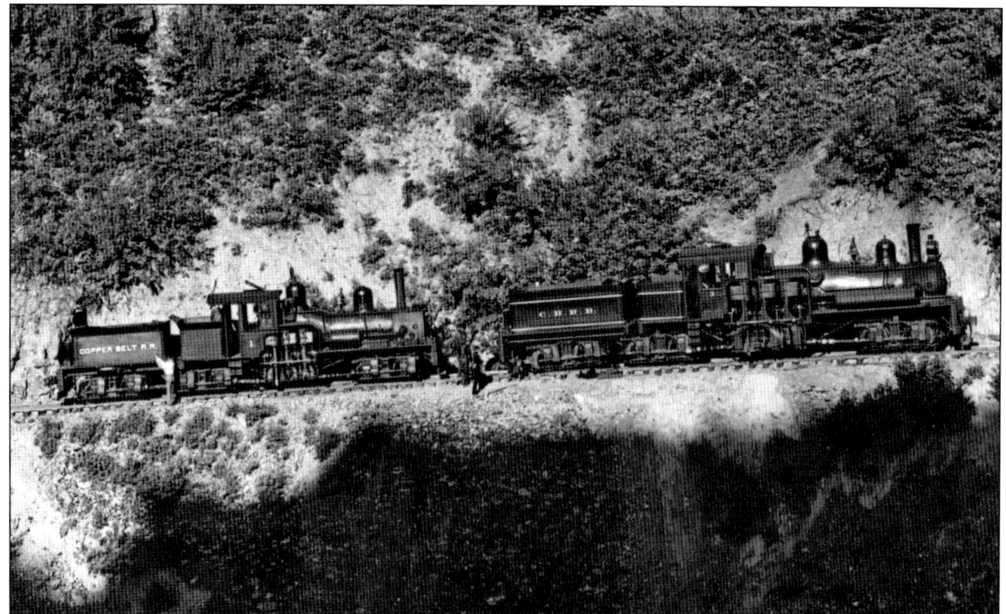

This image shows two Shay locomotives on the Copper Belt Railroad on one of the steep grades at Bingham. The three-mile line consisted of mostly 3.7-percent grades, with the lower portion at 7 percent. The steepest grade was the spur line to the Commercial Mine at 7.4-percent grade, thus the need for Shay locomotives. The Copper Belt Railroad had 34 to 40 forty-degree curves and used 52-pound rail.

The diagram shows the Highland Boy aerial tramway lower terminal station. Another way to get the ore down the mountain was by aerial tramways. Utah Consolidated Mining (Highland Boy Mine) aerial tramway ran from 1897 to 1910. It was about 12,500 feet to the D&RGW station at Frog Town (it was a Bleichert design). In 1910, the aerial tramway changed direction; now it traveled 20,000 feet to the Smelting & Refining Co. smelter in Tooele.

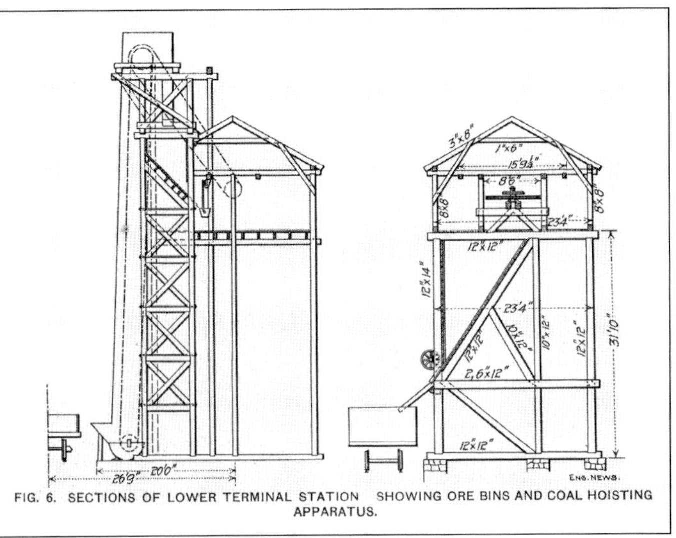

FIG. 6. SECTIONS OF LOWER TERMINAL STATION SHOWING ORE BINS AND COAL HOISTING APPARATUS.

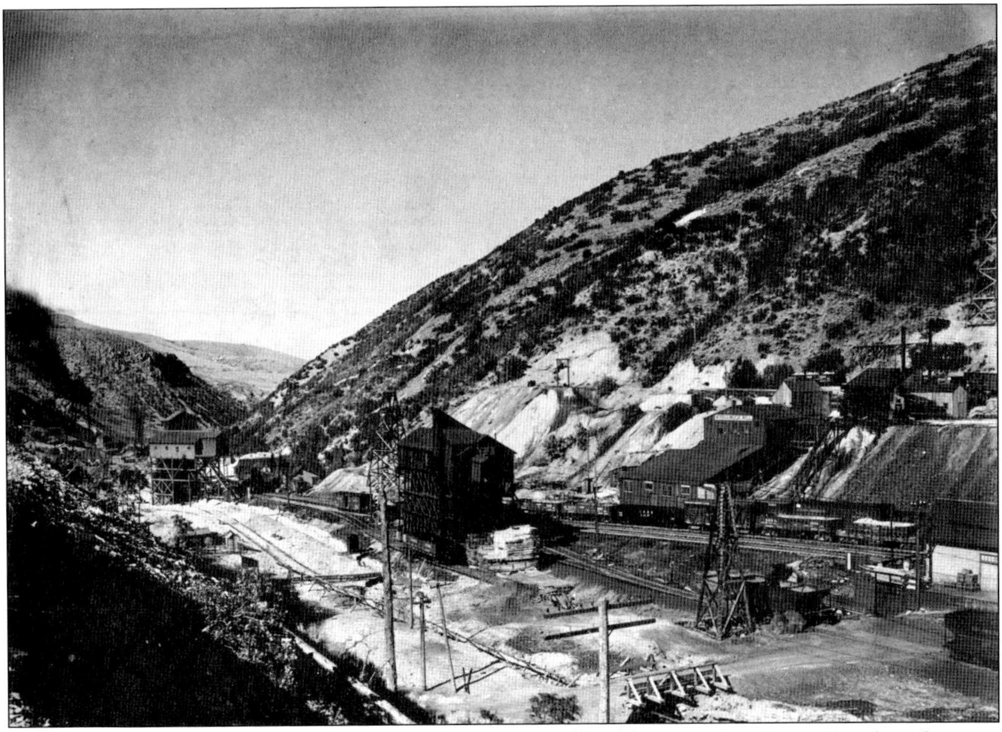

This image shows the large lower aerial tramway terminal buildings in Frog Town. Bingham became home to four aerial tramways, and this photograph shows two: the Utah Consolidated Mining (Highland Boy Mine), center, and the US Mining Company aerial tramway, center left. In view is the Denver & Rio Grande Western depot, bottom right, as well as the Winnamuck Mine and mill. From here, the ore would travel to the valley smelters by railcar.

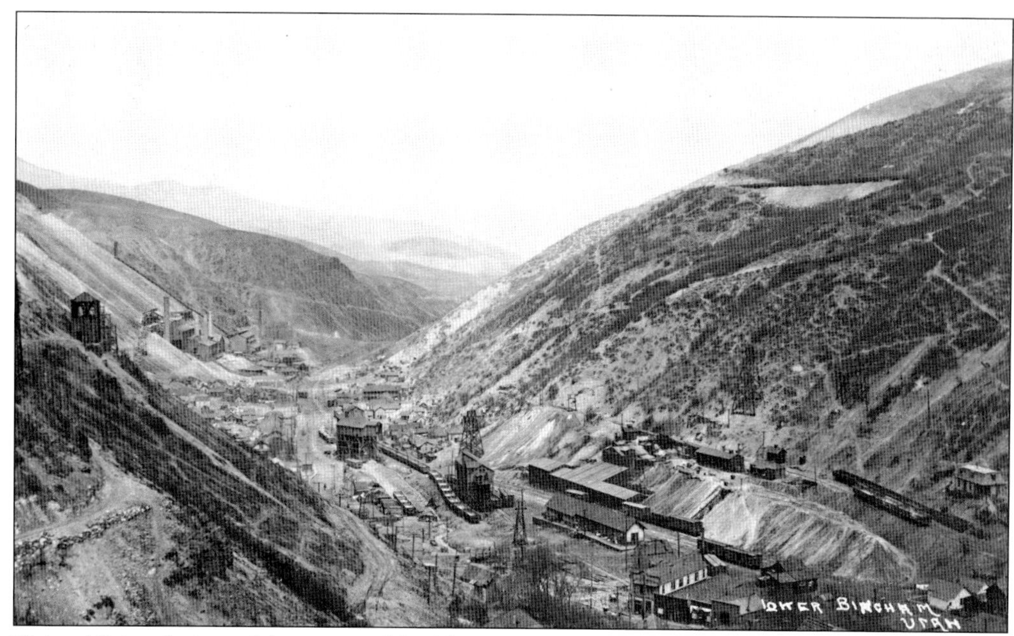

This additional view of the terminal buildings in Frog Town shows a third aerial tramway that went into the Yampa smelter, center left. The Yampa smelters' life was short, only operating from 1904 until 1910. The first ores were brought down from the Yampa Mine to the smelter in Frog Town via the Copper Belt Railroad. The railroad struggled to keep up, so an aerial tramway was built in 1908.

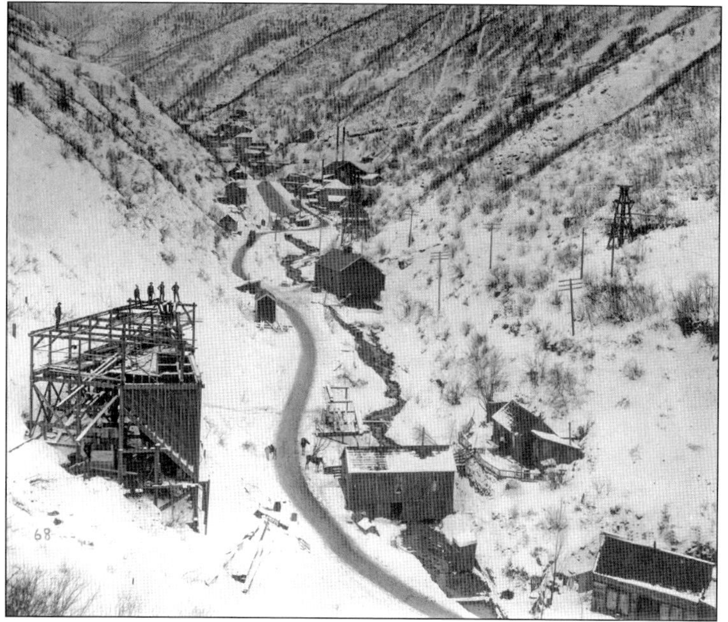

This construction crew for the Utah-Apex Mine works on an aerial tramway building in lower Carr Fork. In the 1870s, Adolf Bleichert, a German engineer, designed the bi-cable aerial tramway. The bi-cable aerial tramway is a stationary cable that supports a series of buckets that are moved by a second moving cable by a hoist. With Bleichert's many patents, the Adolf Bleichert and Company was established to design and install aerial tramways worldwide.

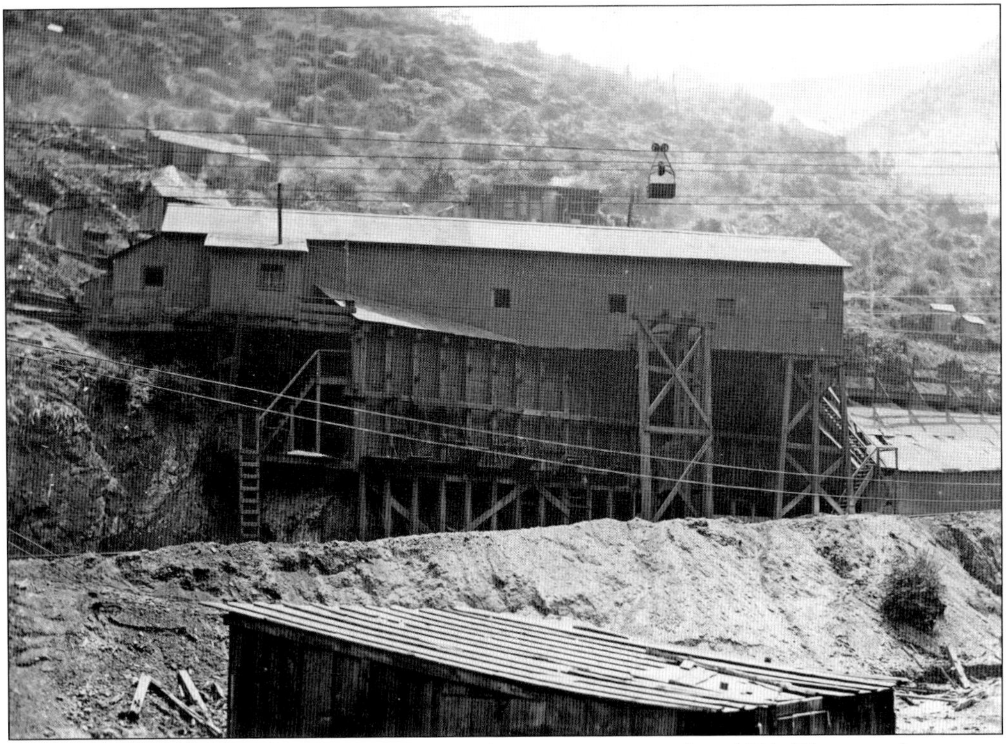

This image shows the Utah-Apex lower aerial tramway terminal in 1914. This aerial tramway transported milled ore from the Apex mill down to ore bins in lower Carr Fork next to a Bingham & Garfield Railway spur line. This aerial tramway was shut down in 1917 when a new underground hoist was installed on the Bingham & Garfield spur line. In 1919, Utah-Apex was sold to Utah Consolidated Mines.

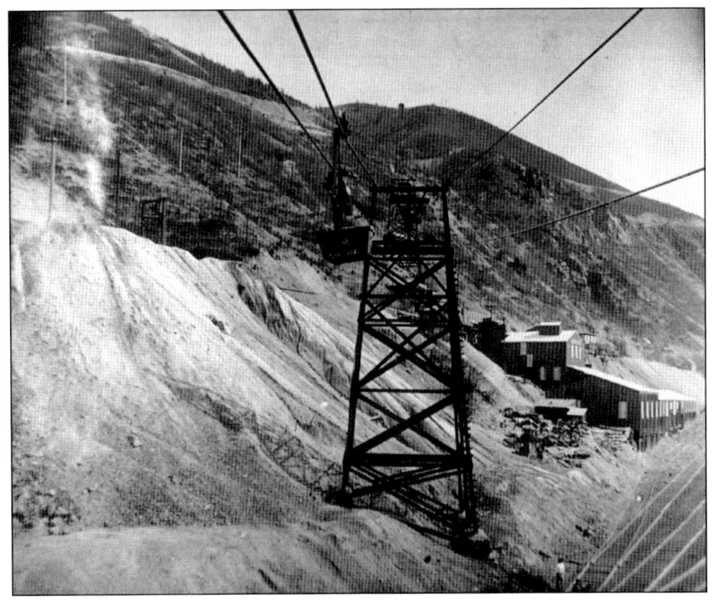

This is an aerial tramway tower for the aerial tramway running to the US Mining Company Mines in Upper Bingham. The distance from the railroad ore bins at Frog Town to the Upper Bingham headhouse was 11,000 feet in length. The US Mining Company's aerial tramway had three branches: the main one, the branch from the Old Jordan Mine to the Upper Bingham headhouse, and a short branch line from the Galena Mine to the Old Jordan Mine.

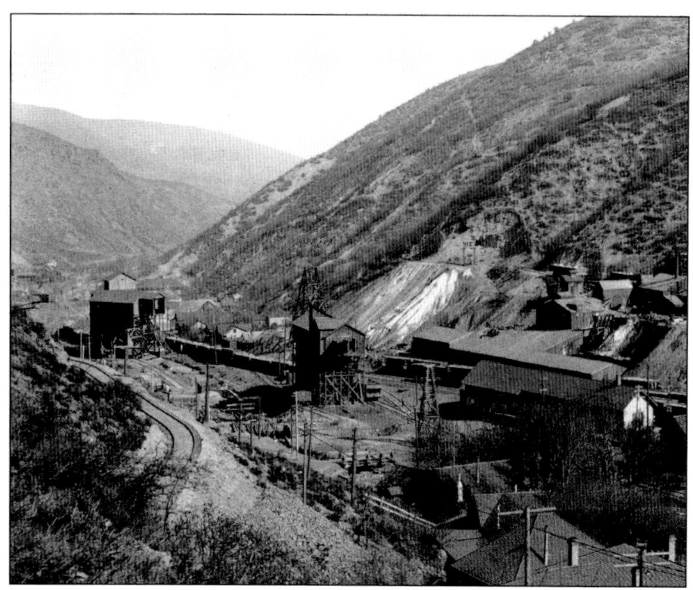

Seen here is another view of the large aerial tramway terminal buildings in Frog Town. The Utah Consolidated (Highland Boy Mine) building is in the center. The US Mining Company aerial tramway building is seen on the left. The Copper Belt spur line track to the Yampa smelter is seen at left. The Highland Boy tramway was equipped with 115 buckets moving 275 tons of ore per day. The US Mining Company aerial tramway speed was five feet per second.

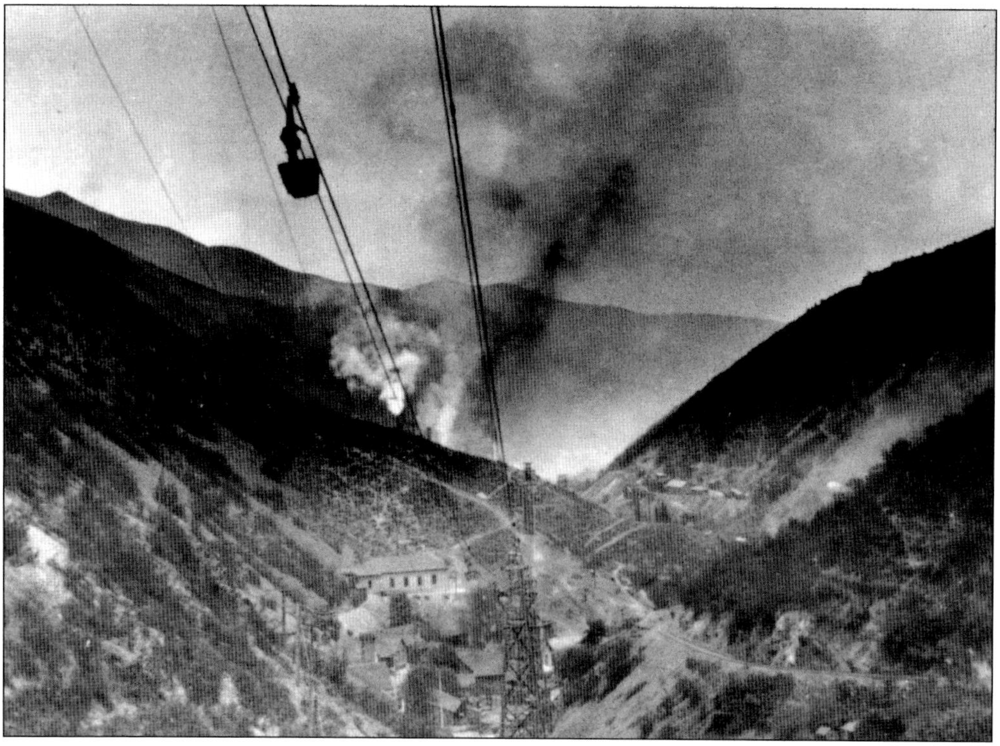

This image shows an aerial tramway (with bucket) looking down Carr Fork Canyon. Smoke from the Yampa smelter fills the canyon, center left. This photograph shows how the ore bucket is rolling along the top stationary cable while the lower cable is moving it along. The aerial tramway also hauled coal up the canyon to their respective mines. From the beginning, miners caught a free ride up and down the mountain on the tramways.

Three

Beginning Open-Cut Mining

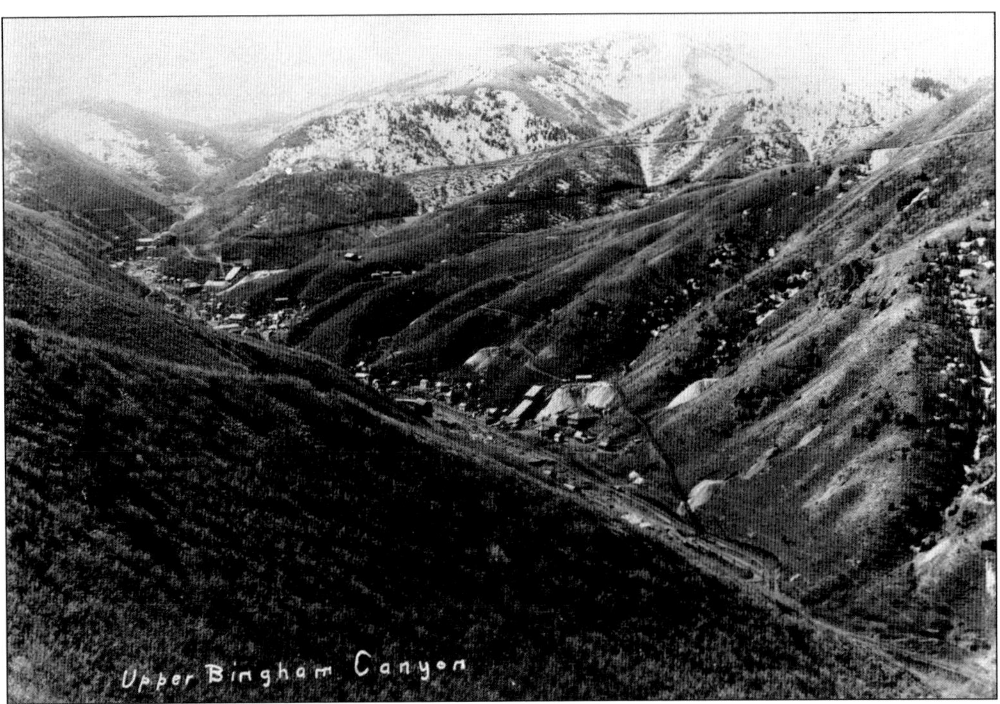

This important photograph, taken in 1904, shows the mountain to the right, which is where open-cut mining started in 1906. Underground mines fill the mountain, located by their visible mine dumps. The canyon floor has a scattering of buildings. The large structures with slanted roofs are mill buildings, where the ore is ground down. One hundred and eighteen years of open-cut mining removed the mountain, and now it is a massive pit.

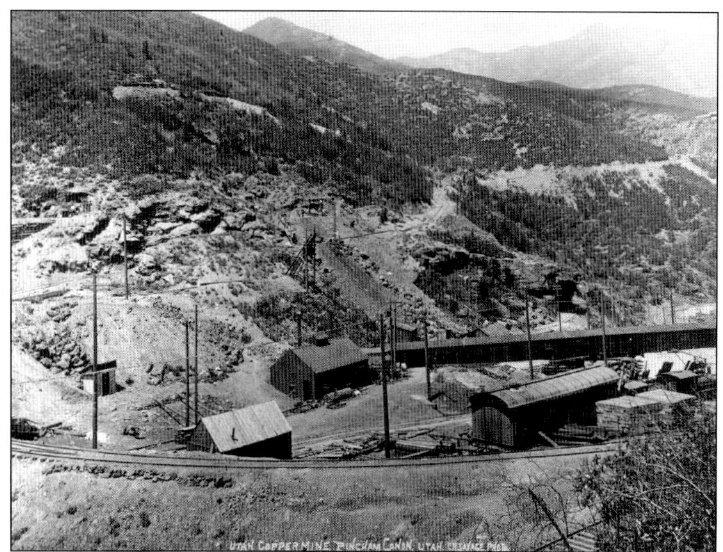

This image shows Utah Copper's underground test mine at Bingham Canyon. It was Daniel Jackling's idea to mass-produce low-grade copper ore by using large steam shovels and trains—this was the first time this method was used with copper. First, Jackling had to assess the feasibility of seeing if this could be done, and he did so by digging this test mine and building a test mill, the Copperton Mill at Dry Fork.

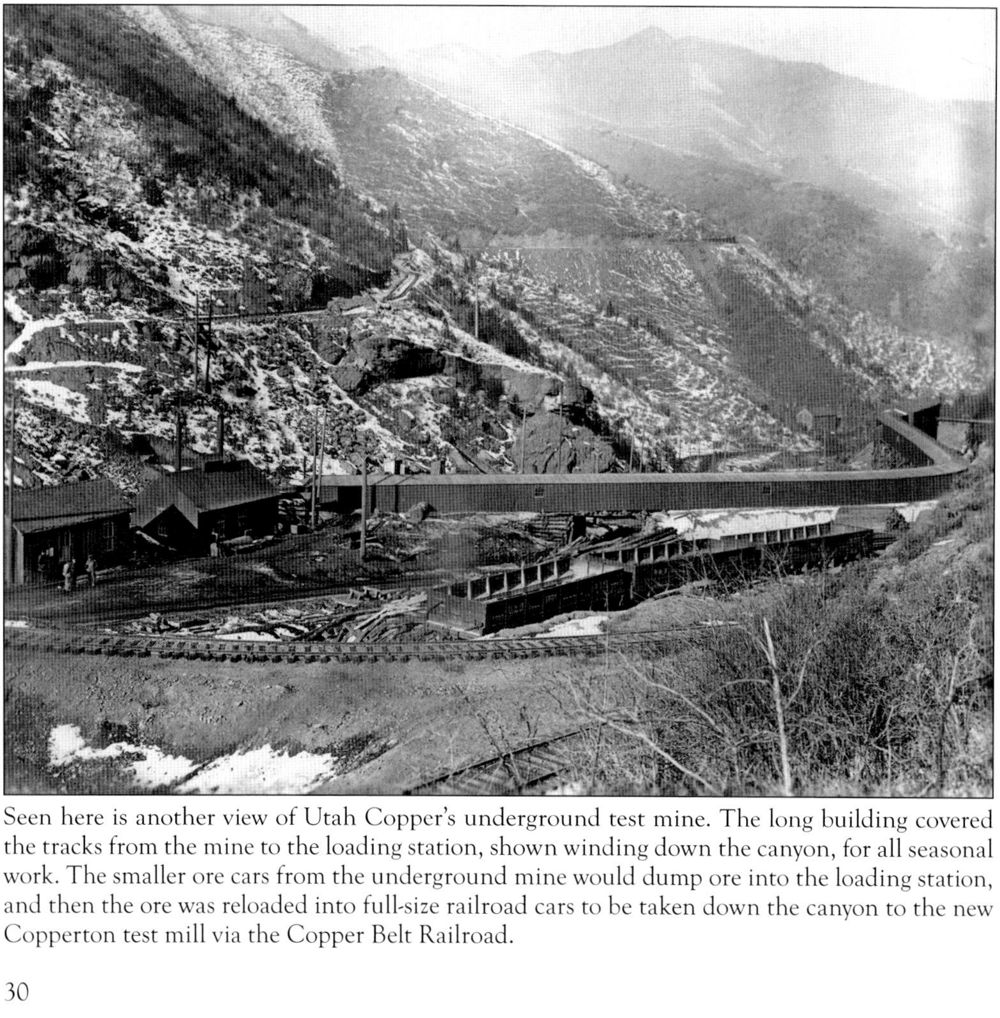

Seen here is another view of Utah Copper's underground test mine. The long building covered the tracks from the mine to the loading station, shown winding down the canyon, for all seasonal work. The smaller ore cars from the underground mine would dump ore into the loading station, and then the ore was reloaded into full-size railroad cars to be taken down the canyon to the new Copperton test mill via the Copper Belt Railroad.

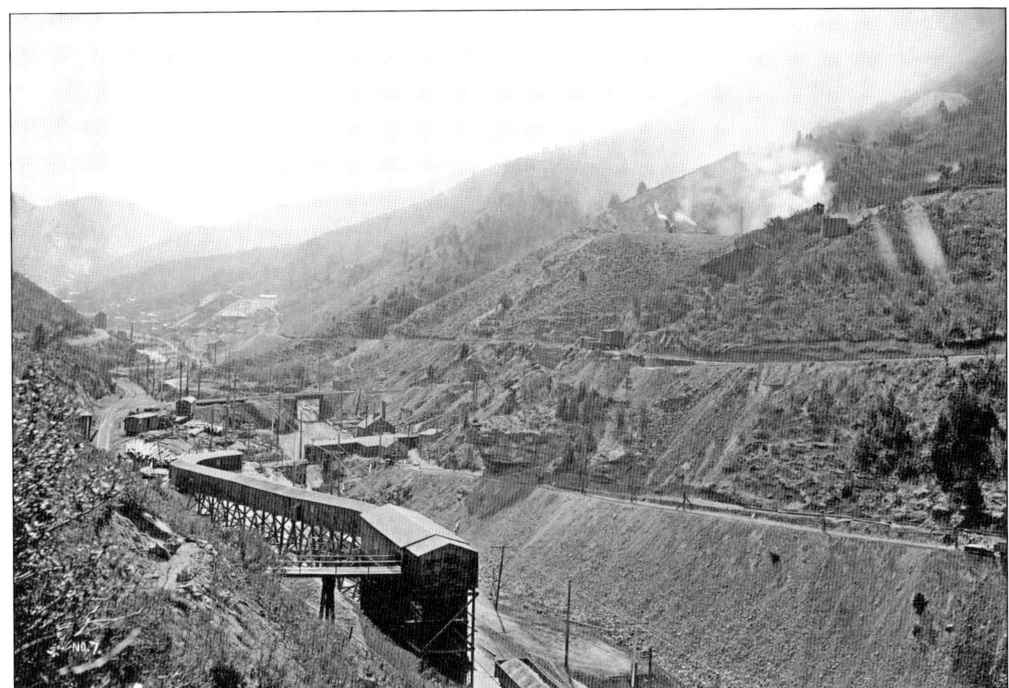

In this October 5, 1907, image, the Utah Copper Company digs away at the mountain, which has been going on for a year. Steam shovels and trains are removing waste rock (overburden); digging started on C-level and D-level, pictured top right. The elevation of the underground mine is 6,340 feet and would be called A-level, which was at bottom of the open-cut mine at center left. Ore is still being mined using the underground mine; this photograph shows the loading station and the covered track that goes to the mine's entrance.

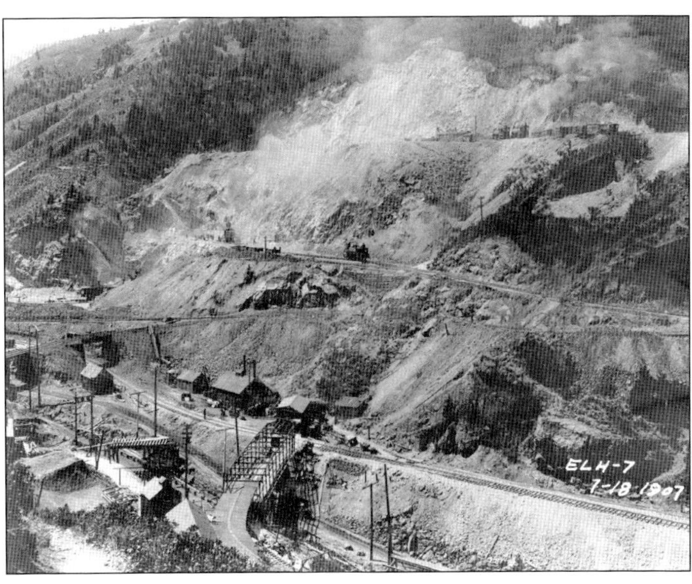

This image, taken July 18, 1907, shows a covered track building and bridge leading to Utah Copper's underground test mine. In the mine, Utah Copper used stope mining, taking out large swaths of low-grade ore at one time instead of following ore veins. In January 1907, Utah Copper stopped underground mining; it had enough ore blocked out to last several years. Moving material with steam shovels and trains was working efficiently.

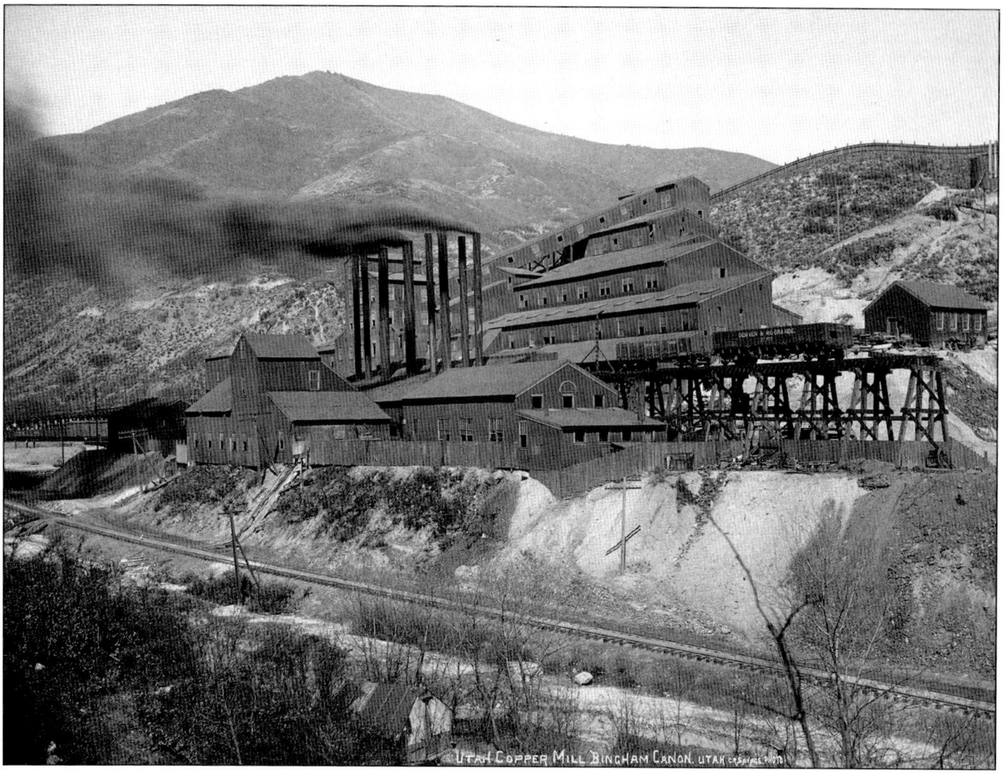

The Copperton test mill, on the corner of Dry Fork Canyon and Lower Bingham, was placed in operation on July 1, 1904. Ore was reloaded into railroad cars at the mine loading station. The Copper Belt Railroad transported the ore to the Copperton Mill. Trains entered the mill over a trestle, and bottom dump ore cars unloaded into a 1,000-ton bin beneath the track.

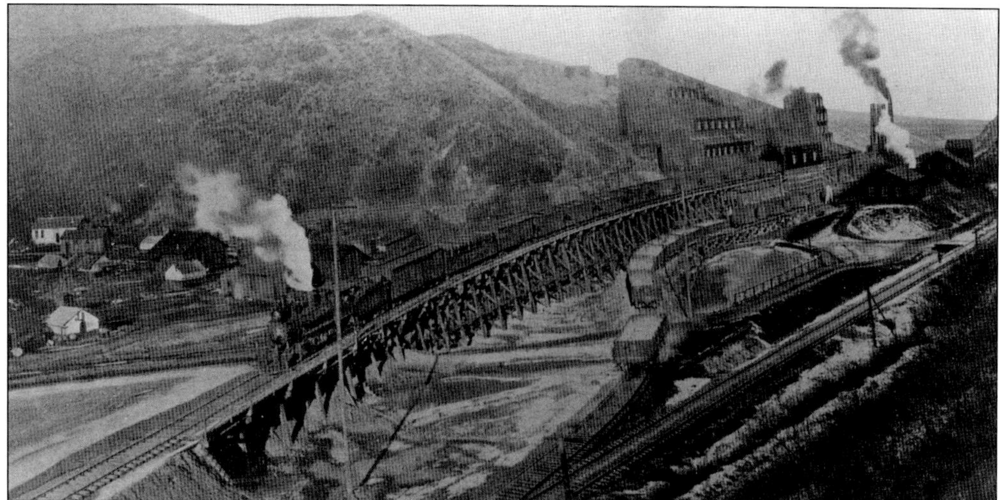

This is another view of the Copperton test mill at Dry Fork Canyon, left. After the start-up of the mill, the Copper Belt Railroad became a very busy line. Utah Copper was sending 500 to 750 tons of low-grade ore per day. In January 1905, the Denver & Rio Grande (not RGW) took control of the line. The concentrate from the copper mill was smelted at the Bingham Consolidated smelter in Bingham Junction (Midvale).

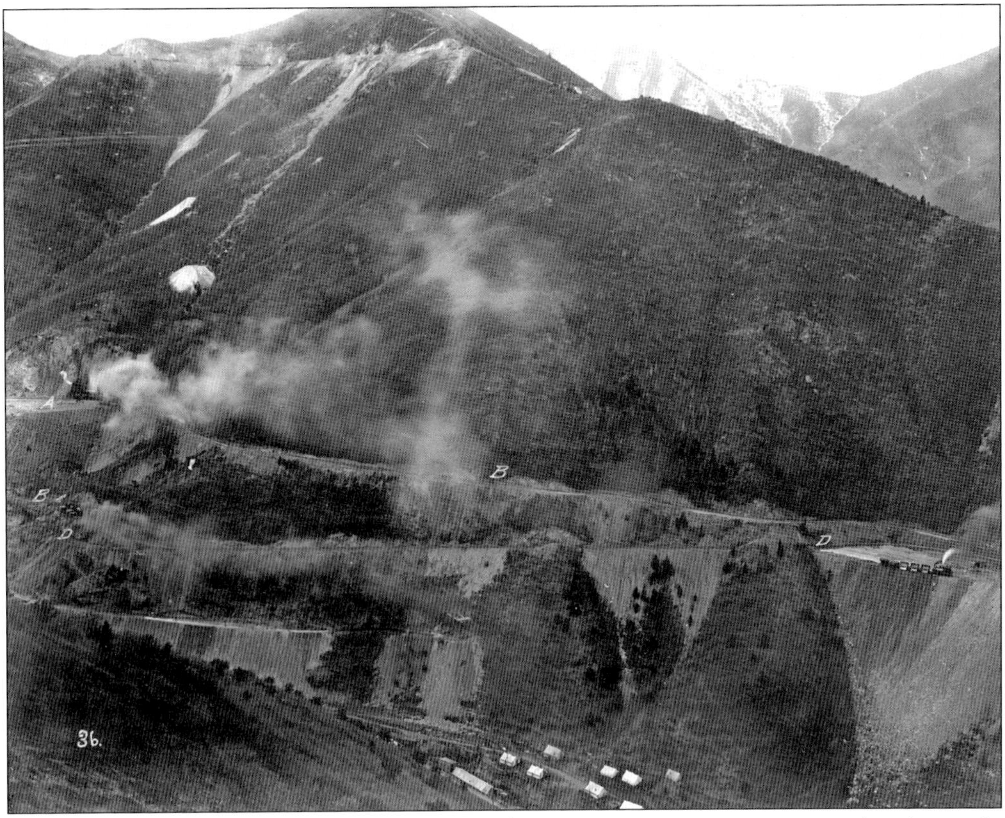

This November 1, 1906, photograph shows a Utah Copper open-cut mining steam shovel at work, center left. To the right, a five-car waste train is dumping overburden. It was Daniel Jackling's idea to transfer waste rock or overburden to the closest side canyons and gulches, which was to save on the expensive start-up costs. The problem is that some old dump sites had to be moved repeatedly. Note the tents seen bottom center.

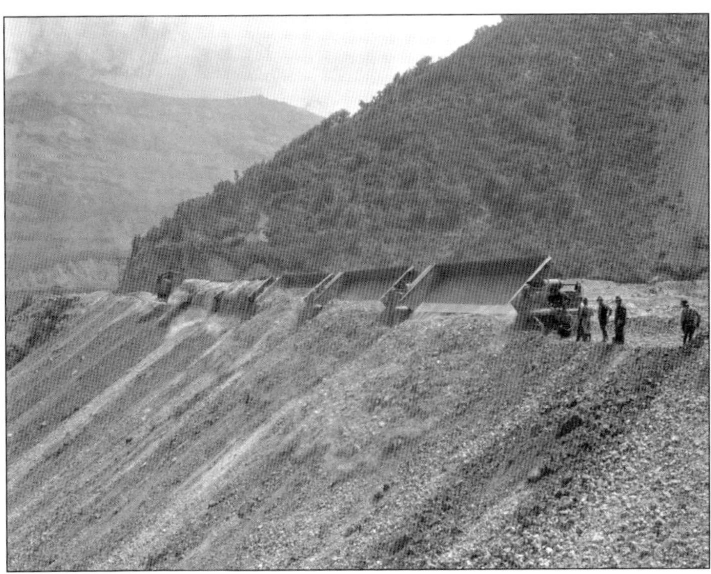

This June 27, 1926, image shows a waste train dumping overburden or capping material from the open-cut mining operation. Though the stripping ratio has changed over the years, the most common is "three to one," meaning three tons of overburden must be removed to get to the one ton of ore. Note that the dump railroad cars are now made of steel and not wood. The dump track gang is pictured center right.

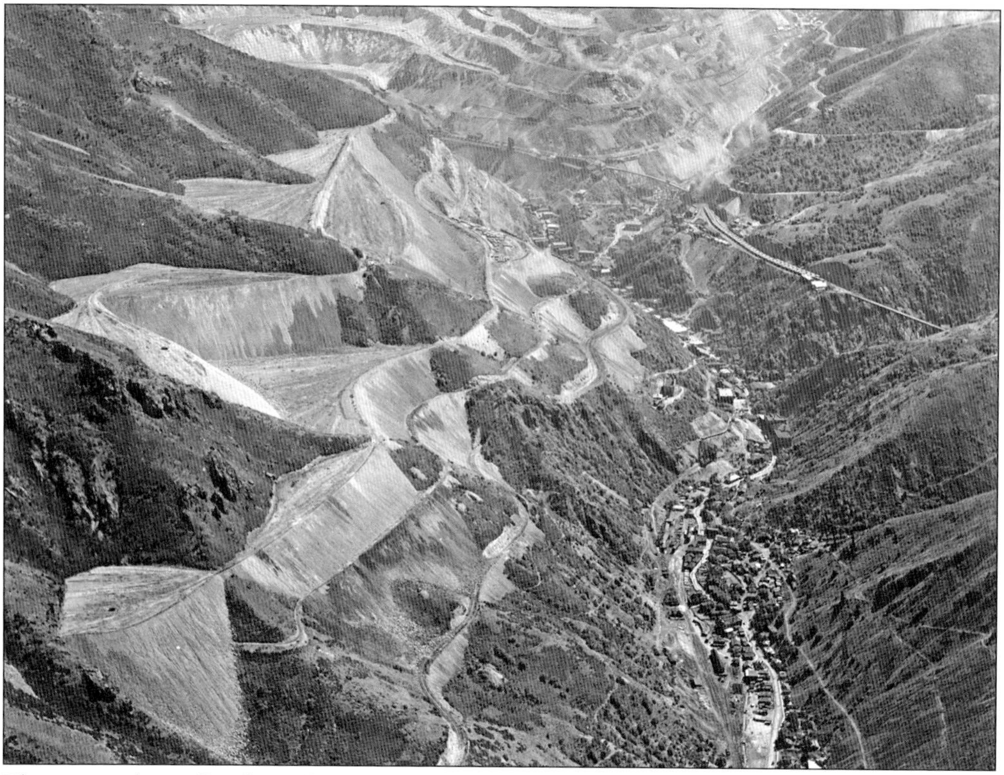

This image shows Bingham Canyon in 1925. The open-cut mine is seen top left, and the side canyons and gulches down the canyon are being filled with overburden. The large Markham and Carr Fork bridges on the Bingham & Garfield Railroad are to the right. The town of Bingham is nestled on the canyon floor. This vantage point shows the confluence where the road splits right to Highland Boy and left for Copperfield.

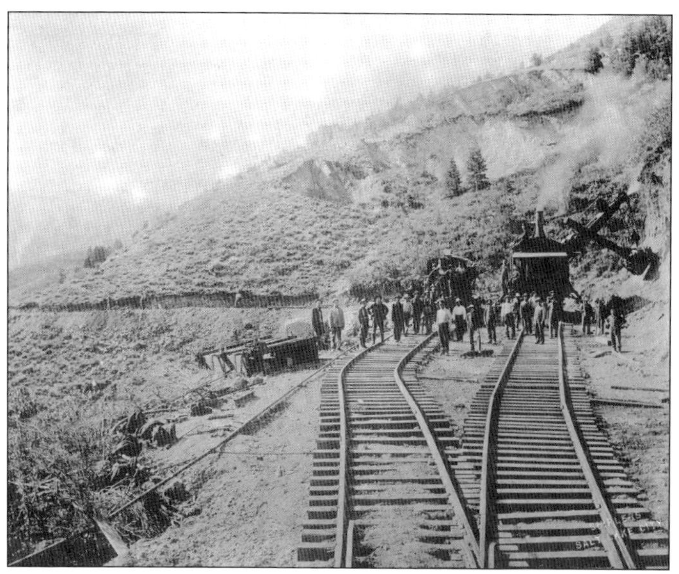

In this photograph, taken on March 16, 1906, Boston Consolidated Mining Company is open-cut mining. Boston Consolidated started its open-cut mining on June 24, 1906, two months before Utah Copper. Boston Consolidated Mining Company was high above on the same mountain as Utah Copper. Note the caved-in mountain above the shovel and men. Boston Consolidated also started with underground mining. This photograph shows the underground mine's collapse.

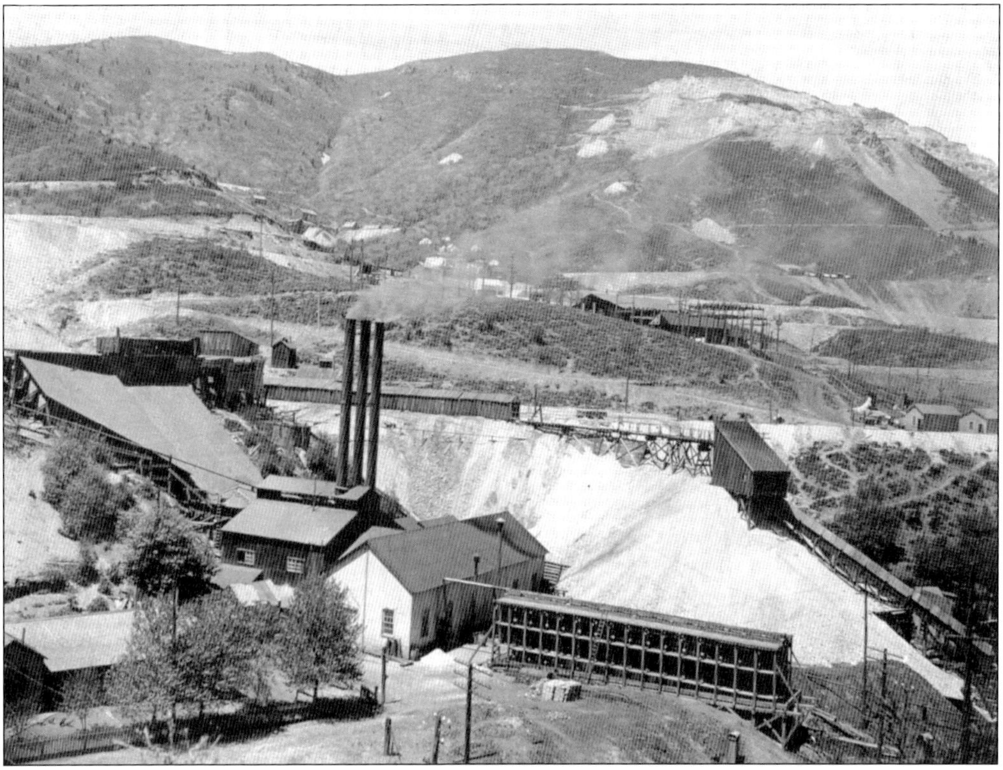

This image shows the Niagara Mine's compressor plant in Upper Bingham with its three smokestacks. In the background, Boston Consolidated Mining Company is open-cut mining high on the mountain, seen top right. Samuel Newhouse, with the sale of his Highland Boy Mine to Standard Oil Company in 1899, purchased a group of claims high on the mountain above Highland Boy covering 350 acres, forming the Boston Consolidated Mining Company.

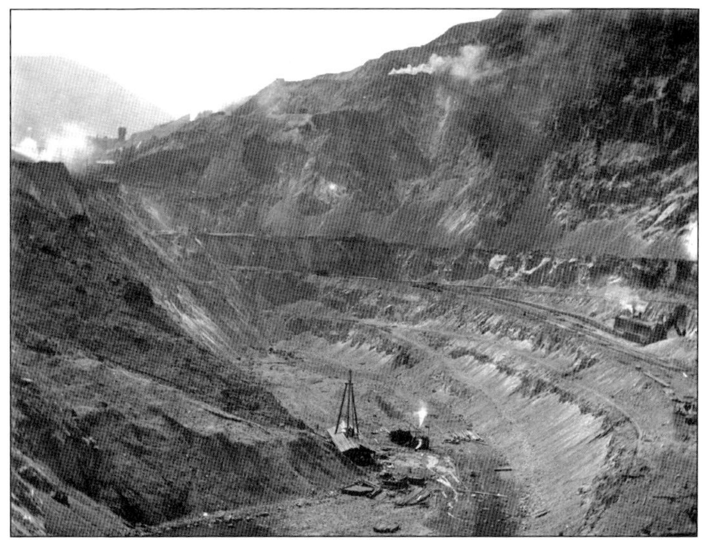

This image, taken in 1914, shows Utah Copper digging away at the mountain. When Utah Copper started back in 1906, skeptics said it could not be done. The company had to attack this huge mountain, building miles of track. Back in April 1906, Daniel Jackling and Robert Gemmell went to Minnesota to the Mesabi iron mines to see their open-cut mining methods, particularly the use of steam shovels and trains.

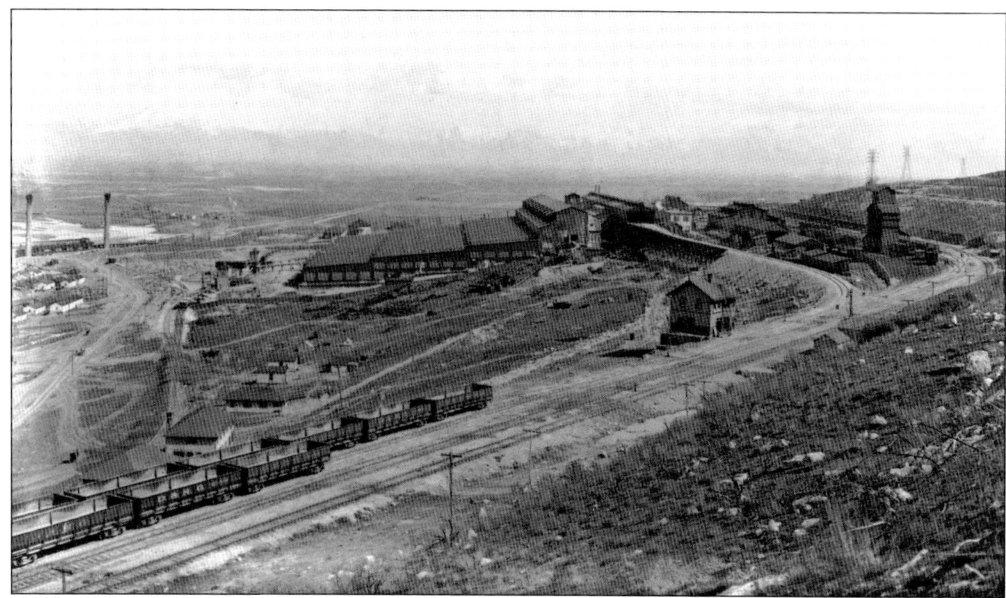

This photograph shows the Magna mill, located at the northern end of the Oquirrh range. With the success of the Copperton test mill, a new, larger mill was needed, with a better water supply and more room for tailings. The new mill's size was 509 by 600 feet. It could produce 6,000 tons of concentrate a day, and the cost was $4,005,000. The Guggenheims helped finance the project.

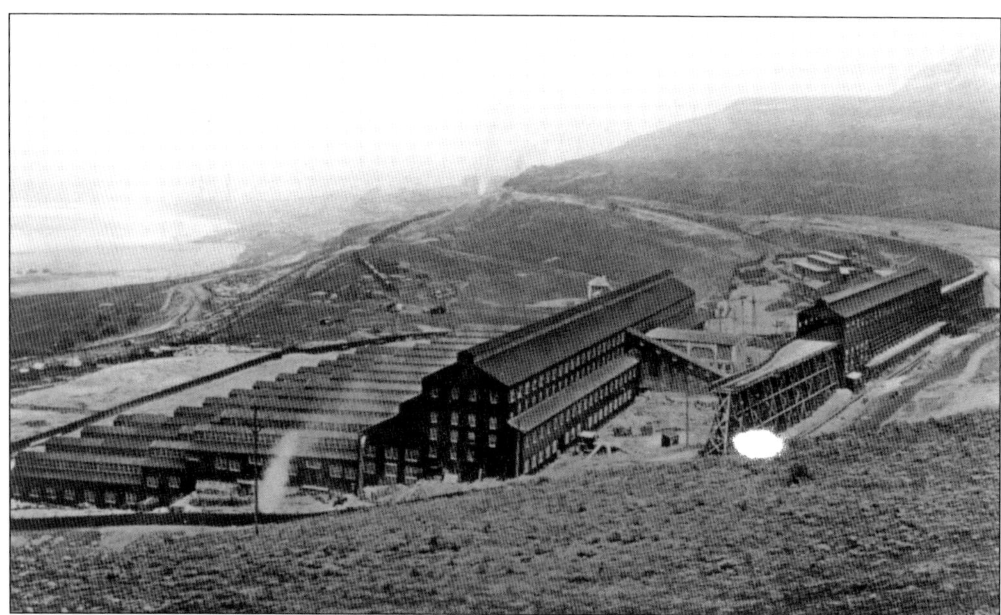

Boston Consolidated Mining Company built its own mill. Samuel Newhouse purchased 910 acres near the Magna mill. Construction started in 1906 and was completed a year later at the cost of $1,468,902; it produced 3,000 tons of concentrate a day. Both Boston Consolidated Mining Company and Utah Copper ore concentrate were smelted at the American Smelting & Refining Company (ASARCO) smelter, just a short distance away. Built in Kessler Canyon, it was the largest copper smelter in the world.

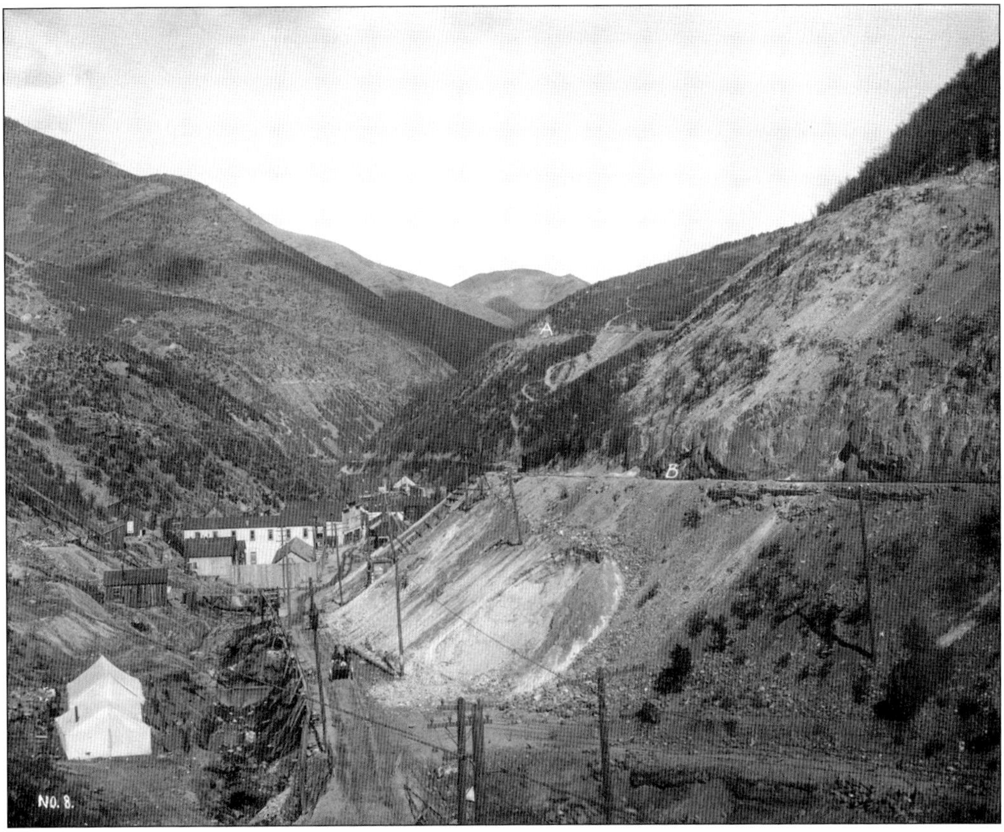

This view looks down Bingham Canyon away from the mine on October 5, 1906. The photograph is labeled with letters. "A" is the Rio Grande Western low-grade line (also called "high line"). Both Utah Copper and Boston Consolidated Mining Company contracted with RGW to move ore from both companies' new open-cut mining operations to their respective mills. "B" is the old Copper Belt Railroad line, now owned by the Denver & Rio Grande.

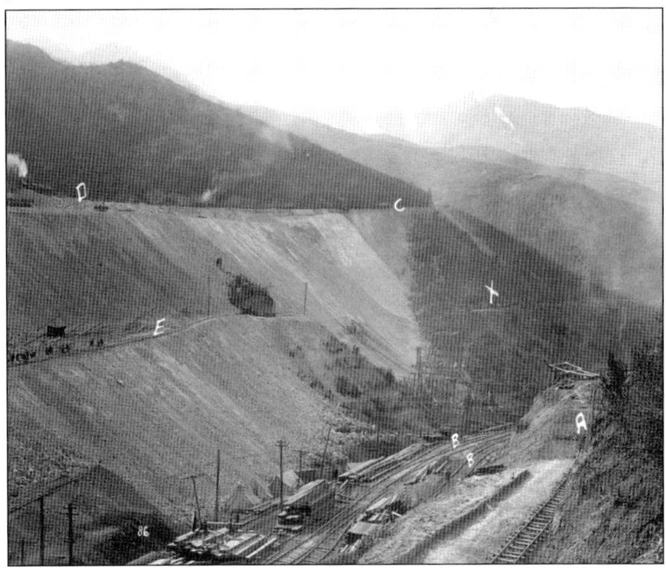

This image, taken April 13, 1907, shows Utah Copper's early open-cut mining. This image is also labeled with letters: the new RGW low-grade line (A), the Copper Belt Railroad (B), the Yampa Mine spur line (C), the Utah Copper steam shovel (D), the auxiliary yard (E), and construction on a new bridge to cross the canyon (X). The auxiliary yard level was called A-level, and its elevation was 6,340; this is where both the RGW and, later, the Bingham & Garfield Railroad entered the mine.

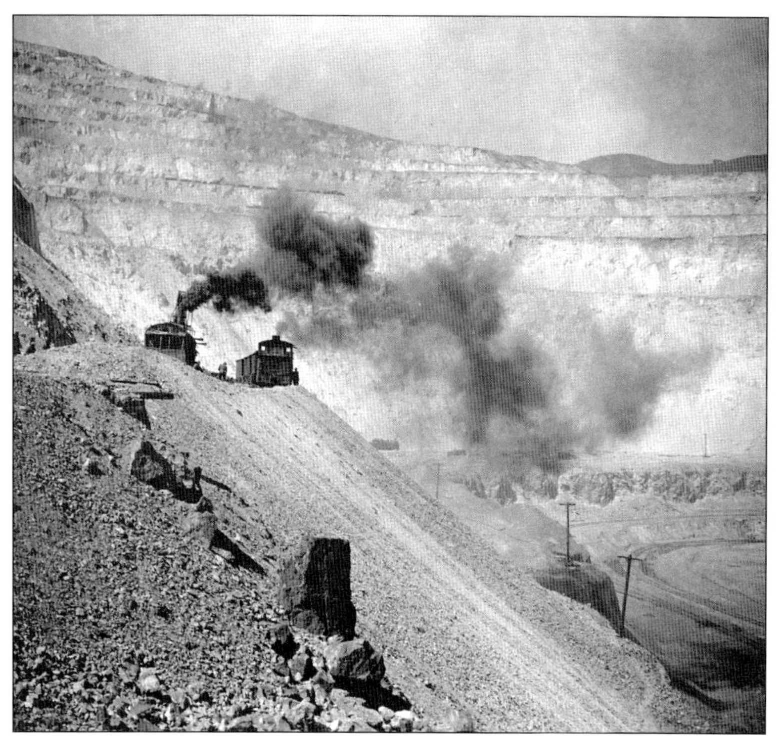

This image shows how steam shovels and trains cut away at the mountain by making workable levels; the most common height was 50 feet. By 1909, Utah Copper had 11 steam shovels in operation with 21 locomotives and 145 dump cars and 16 miles of 65-pound standard-gauge railroad in the mine. Note the black smoke billowing out of the shovel. The steam-powered shovels were converted to electric-powered in the early 1920s.

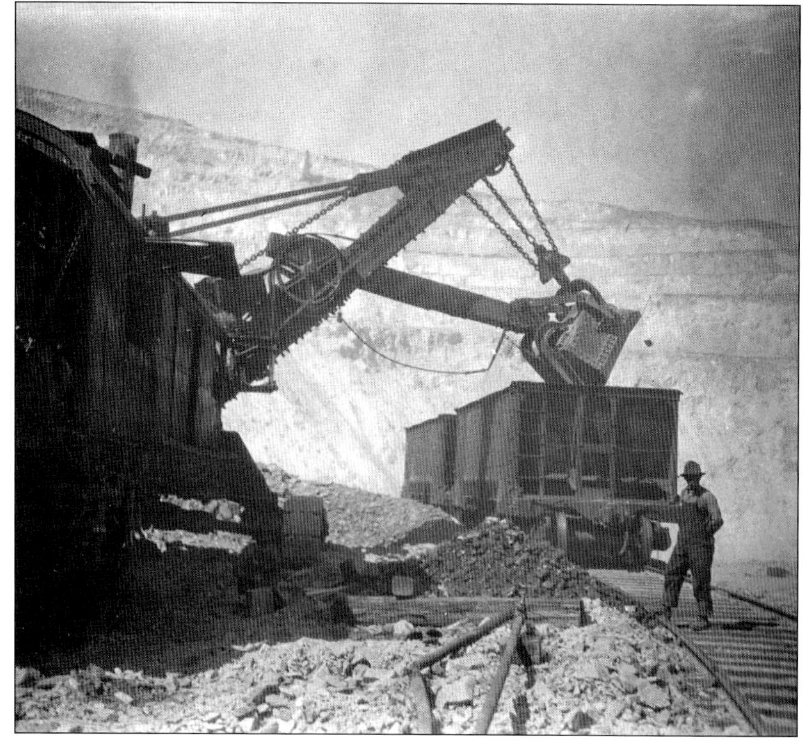

This image shows a steam shovel loading an ore car. There is a hose running to the shovel and smoke coming out of the stack. The man is standing next to a pile of coal, which means the shovel is still steam-powered. After the shovel digs along the level, loading ore or waste cars, the railroad track needs to be moved closer to the level for the next shovel pass.

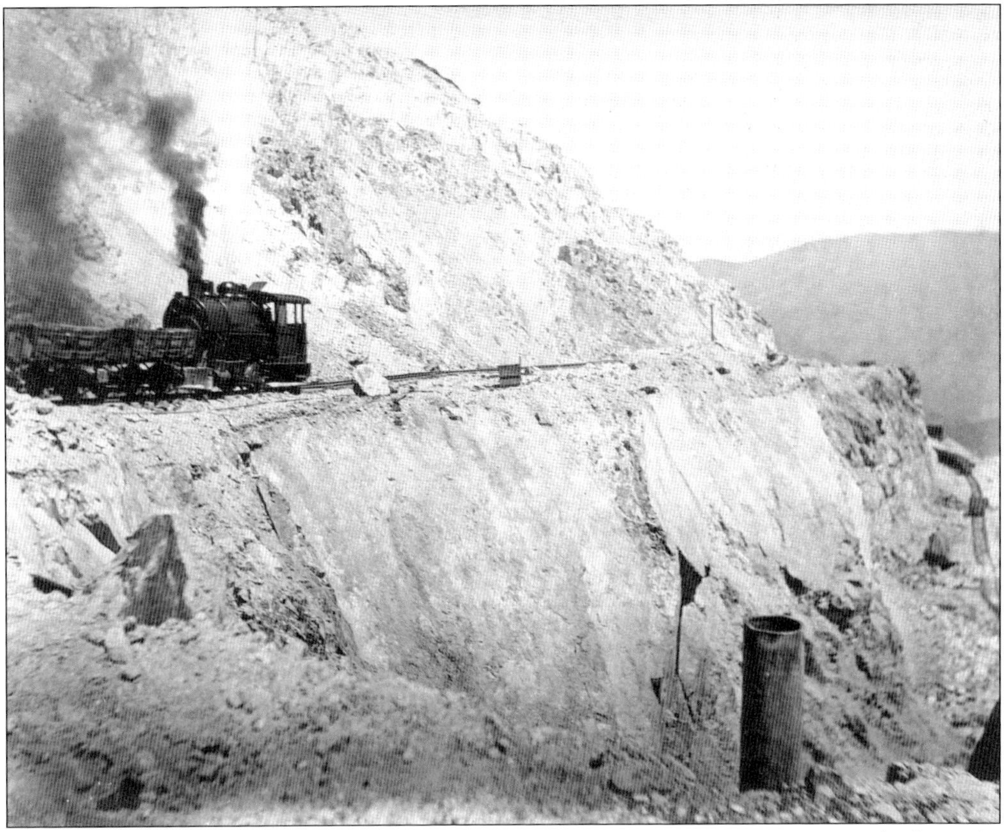

This photograph shows a small Porter steam locomotive and two wooden waste cars high on the mountain. Boston Consolidated Mining Company is also digging away at the top of the mountain. By 1907, Boston Consolidated had five 90-ton steam shovels, four Marion and one Vulcan, all with five-yard dippers, as well as ten 19-ton Porter narrow-gauge locomotives, one 90-ton Shay standard-gauge locomotive, and 200 four-yard dump cars. Boston Consolidated Mining Company used both narrow- and standard-gauge trains.

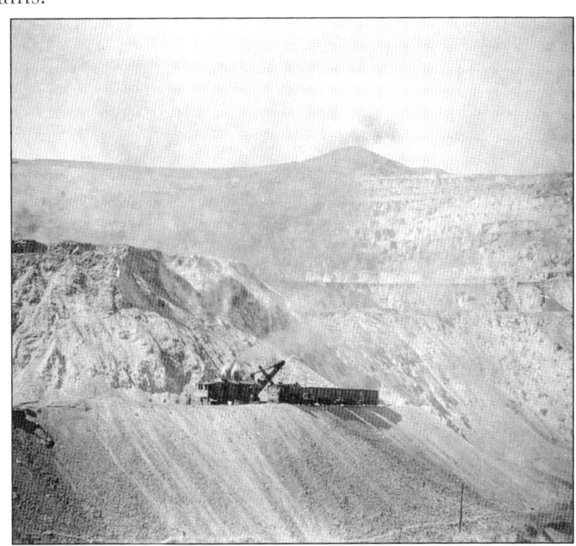

This image shows a steam shovel, a small locomotive, and three ore cars at work. Note the number of workers on the ground, as this method of shovel-to-train mining was labor-intensive. The steam shovels had the operator, oiler, and fireman, while the locomotives had the engineer, brakeman, and fireman. Track gangs had to keep the rails clear of debris and were constantly moving the rail closer to the level.

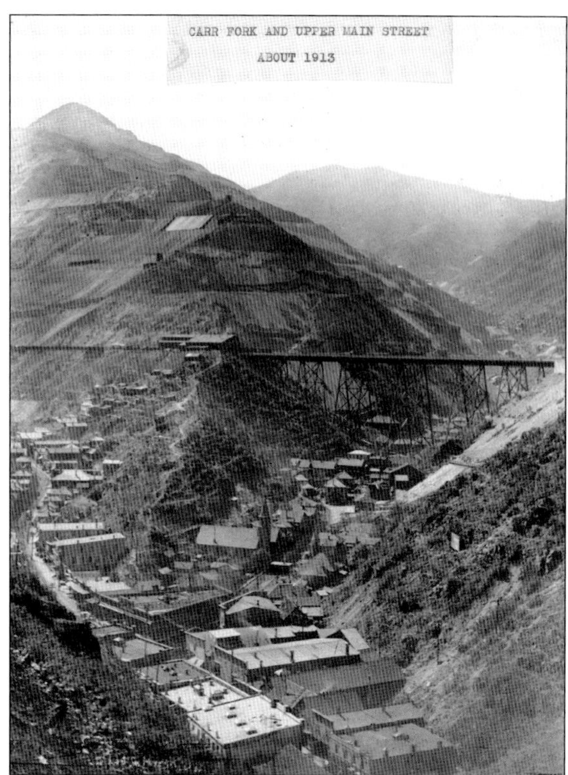

This image shows the confluence where Bingham Canyon is divided, and the mountain where open-cut mining was taking place; note the town of Bingham below. Even when open-cut mining started in 1906, it was evident that Utah Copper and Boston Consolidated Mining Company would merge. With both companies digging away at the same mountain, one would have to be in control. Utah Copper and Boston Consolidated Mining Company merged on March 1, 1910.

The Utah Copper mine is seen here on May 7, 1926. Twenty years of constant digging show how much of the mountain has been removed. The bottom of the mine is A-level (elevation 6,340), seen at the bottom center. From here on up the mountains, the levels or terraces are named letters, like "B," "C," "E," and so on, up the mountain. A-level would be the bottom of the mine for many years.

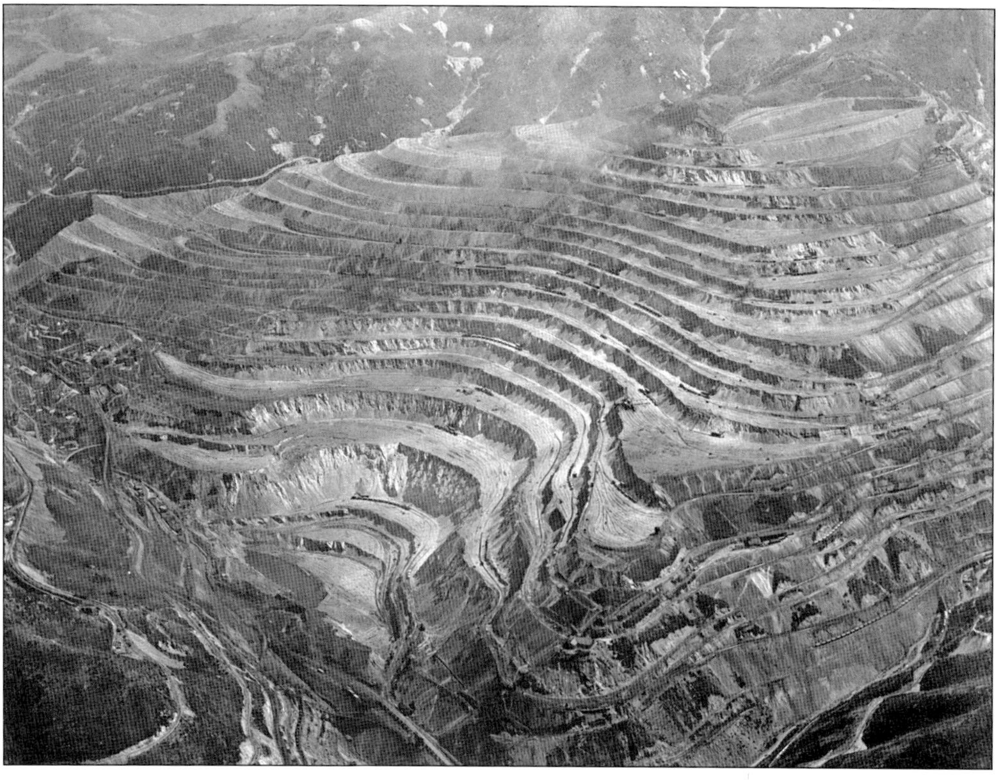

Four

The Steam Era

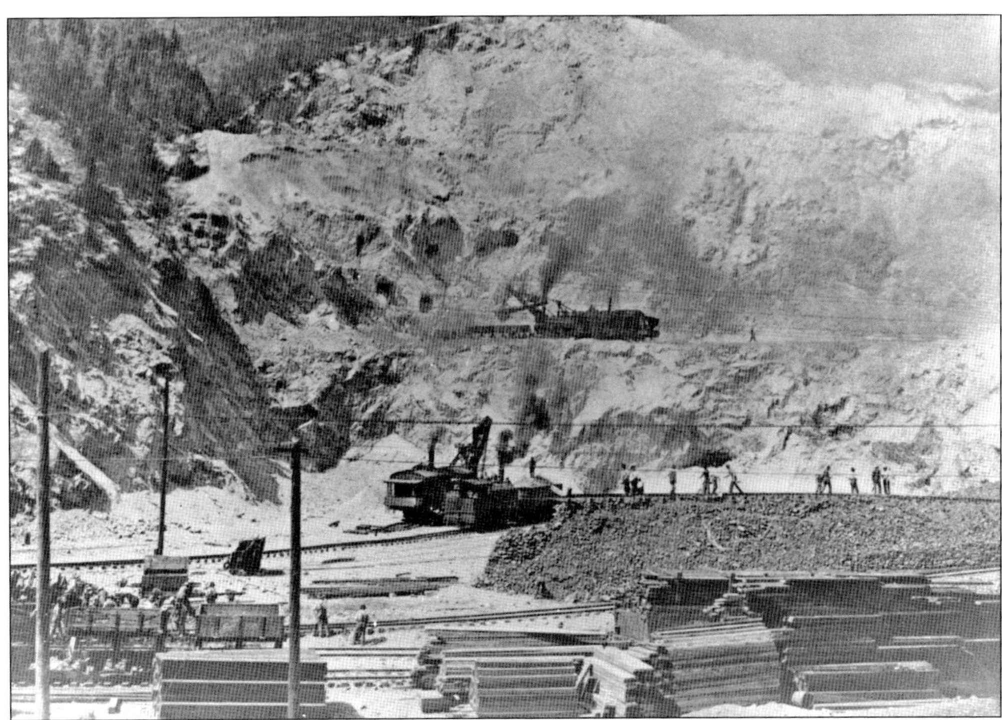

The Utah Copper mine was a busy place in the early years. It was the steam era, and it was only with the use of steam shovels and trains that this work could be done. The images show two steam shovels at work, loading waste cars, and many workers on the ground. Note the wooden waste cars, pictured bottom left. Where the top shovel is digging, old mine shaft holes can be seen.

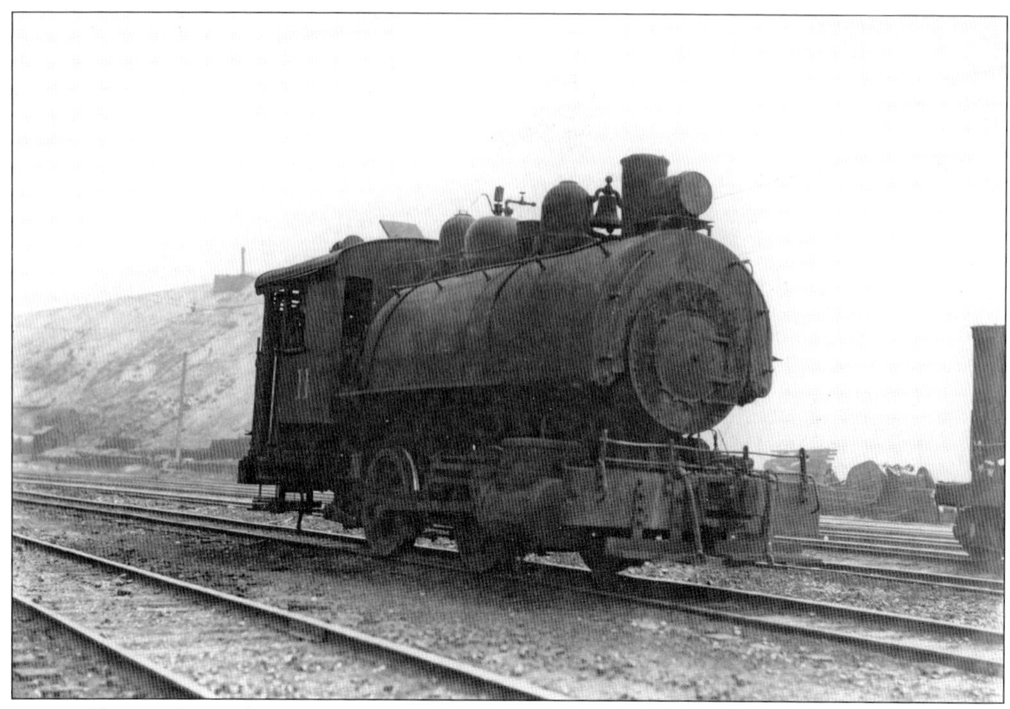

Pictured here is the Utah Copper No. 11 Porter 0-4-0 saddle tank locomotive, builder date June 1907. The first steam locomotives that were used by Utah Copper were Porter and Davenport saddle tank engines, nicknamed "dinkeys." There was a whistle stop on the H-level dump line where these small engines could get water. When houses started being built around the area, it was called "Dinkeyville" after these small steam engines. Utah Copper purchased 21 of these dinkeys from 1907 to 1910.

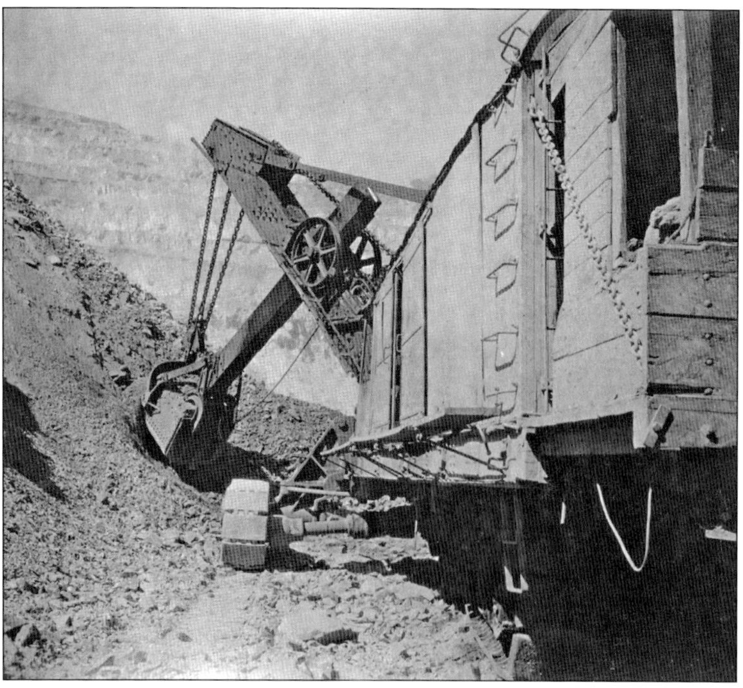

A standard steam shovel is pictured at work in the open-cut mine. The shovel has been upgraded with crawler treads; the first shovels at the mine ran on standard-gauge rails. That required a parallel set of tracks that had to be constantly built and maintained in front of the working shovel. They started to be replaced with crawler treads in the early 1920s. The first shovels were built by Marion, Bucyrus, and Atlantic.

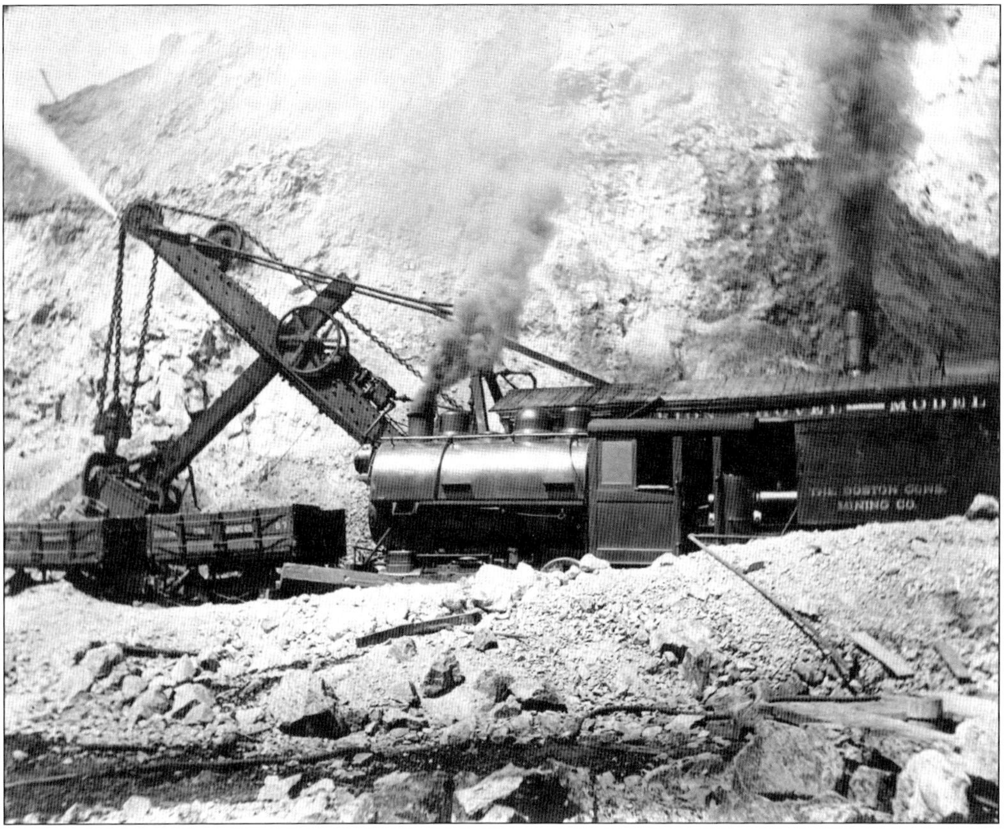

This image shows a Boston Consolidated Mining Company steam shovel loading wooden waste cars. The Boston Consolidated Copper Company property was high on top of the same mountain where Utah Copper was digging. Both were using similar equipment, except Boston Porter locomotives were narrow gauge, which included Nos. 1–9, 11, and 12. No. 10 was a standard-gauge Shay locomotive, and pictured is the No. 11 Porter locomotive, built in April 1907.

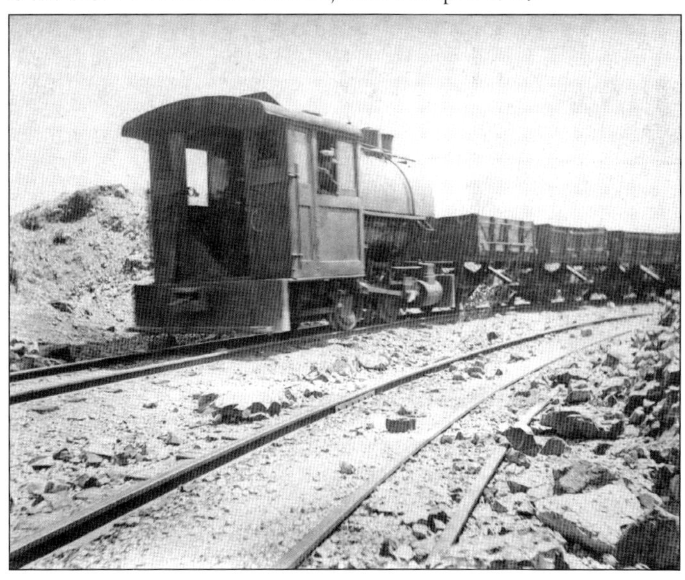

The image shows a waste train working on the Boston Consolidated Mining Company open-cut mine. This photograph shows the three-rail system when using both standard- and narrow-gauge trains. Most commonly, narrow gauge is three feet, compared to the four feet, eight-and-one-half-inch standard gauge. The first ore trains coming from the mine had to work their way down a series of treacherous switchbacks at the top of Carr Fork Canyon.

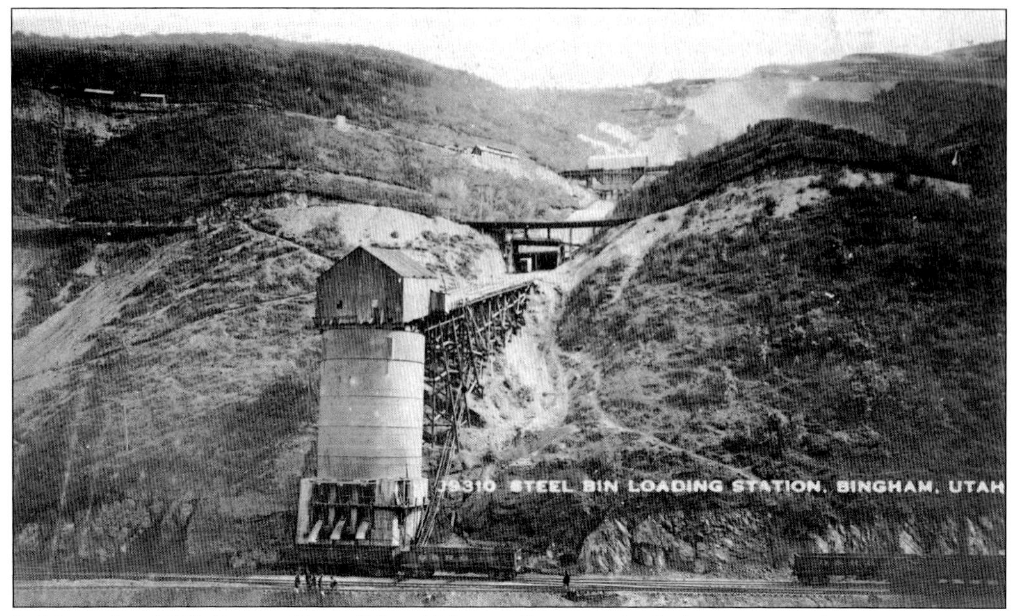

In November 1907, Boston Consolidated Mining Company built this gravity rail tramway to move its ore down the mountain. The tramway was 2,100 feet long with a 27-percent grade. The tramway had a parallel set of tracks with skip cars of 12 tons each, counterbalancing each other as it moved loaded cars down and empty cars back up. The skip cars dumped their ore into a large 3,000-ton steel storage bin at the bottom.

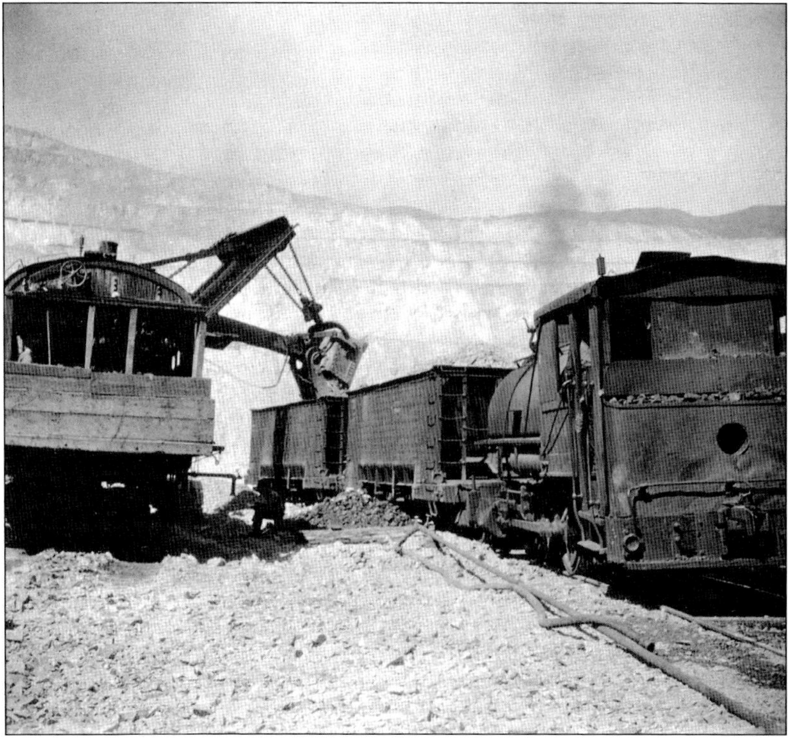

This image shows a steam shovel loading two steel ore cars. Note the coal bin on the back side of the locomotive. To the left of the locomotive on the ground is a large water hose for the steam shovel. There is also a pile of coal on the ground where the man is sitting. It was hard to keep the steam shovels and trains supplied with water and coal.

This image shows the development of Utah Copper as it digs away at the mountain. When open-cut mining started in 1906, Utah Copper used secondhand equipment, two Marion and one Vulcan steam shovels, four small Davenport standard-gauge steam locomotives, and several six-yard wooden dump cars. Notice the condition of the wooden dump cars at the bottom of the photograph—hard rock mining takes its toll on them.

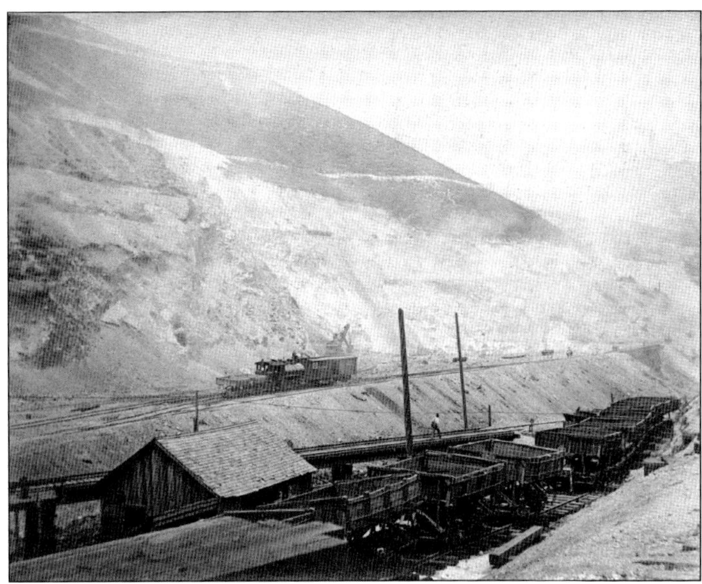

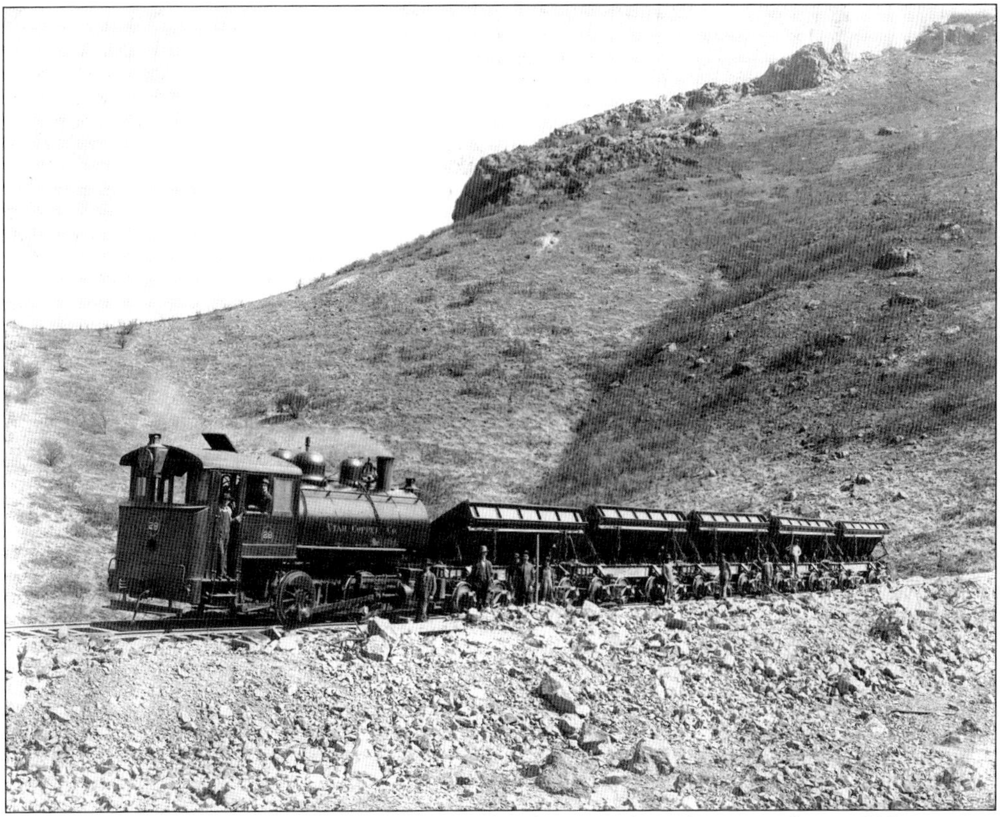

Utah Copper's No. 28 locomotive is dumping overburden, away from the mine. The No. 28 locomotive with its five steel dump cars looks new as 11 men pose for a photograph. The locomotive is a Porter 0-4-0 saddle tank; the builder date is July 1910. By 1914, Utah Copper had 100 standard-gauge 12-yard all-steel dump cars, 117 standard-gauge six-yard wooden dump cars, and 144 narrow-gauge four-yard wooden dump cars.

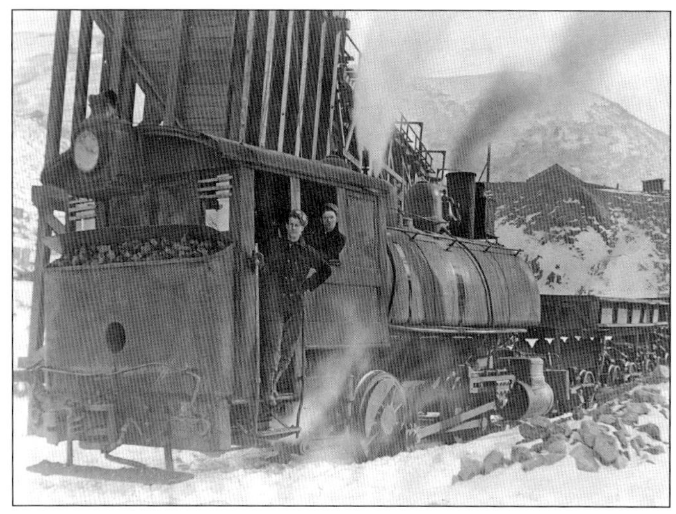

This image shows the Utah Copper No. 21 Porter locomotive with all-steel 12-yard dump cars. What is interesting is the large ore-loading station behind the train, one of many around the area. After the merger with Boston Consolidated Mining Company, Utah Copper kept expanding by buying up old claims (mostly for surface rights) to put overburden or waste rock, using every available side canyon and gulch that was close to the open-cut mining operation.

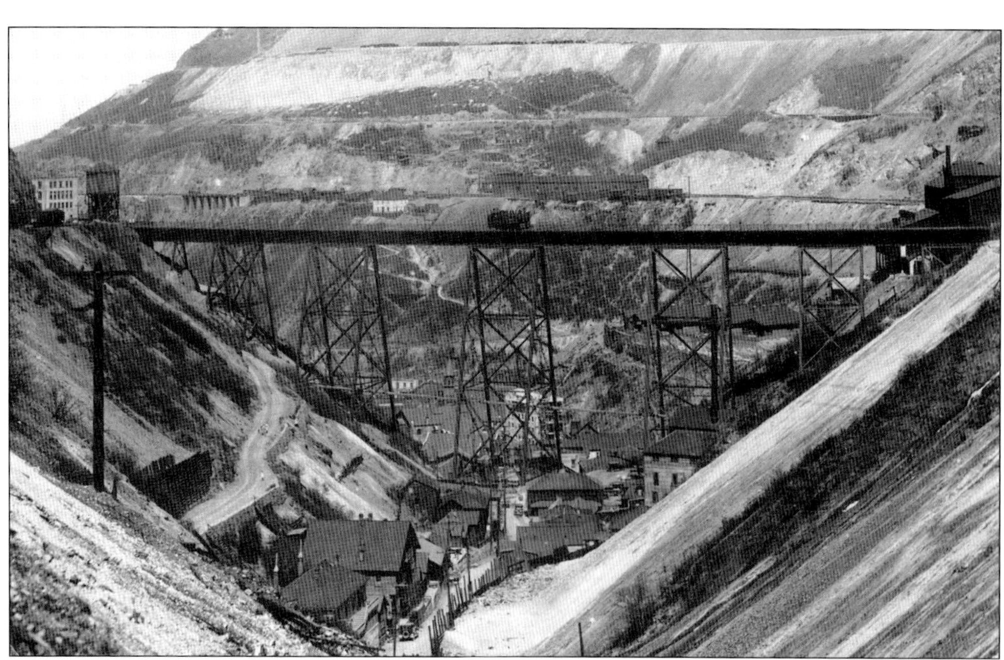

A small, lone, side-tank locomotive is crossing over the large Carr Fork Bridge. Utah Copper had several side-tank locomotives; most were Baldwin 0-6-0 and two Porter 0-4-0 side-tank locomotives. This one is the Utah Copper No. 87 Baldwin 0-6-0 side tank. The Carr Fork Bridge was part of the Bingham & Garfield Railroad. Across the canyon is the old Rio Grande Western low-grade line with its Cuprum yard.

Work was done on the large roster of steam locomotives at Utah Copper's machine shop. It was a long, narrow building perched on the mountainside next to the Carr Fork Bridge. Around the machine shop were the carpenter shop, the oil house, and supply sheds. Note all the tracks next to the buildings; this was the staging yard where ore cars were coupled together into larger trains for the trip to the mills.

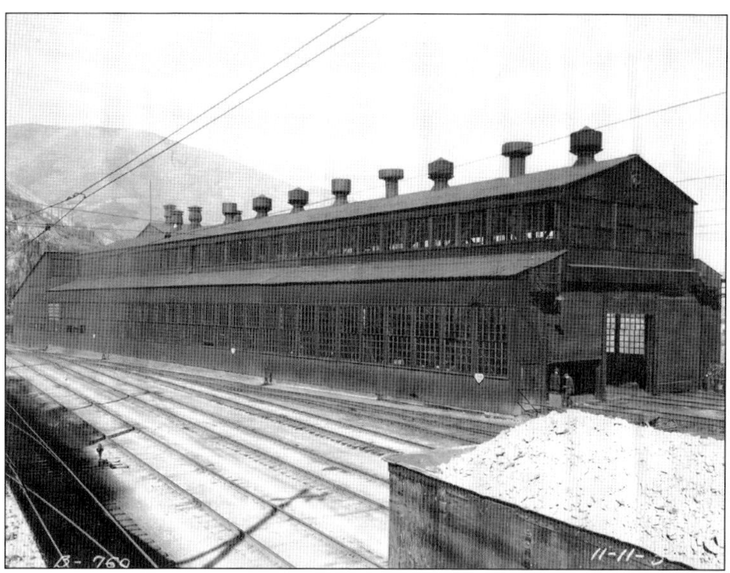

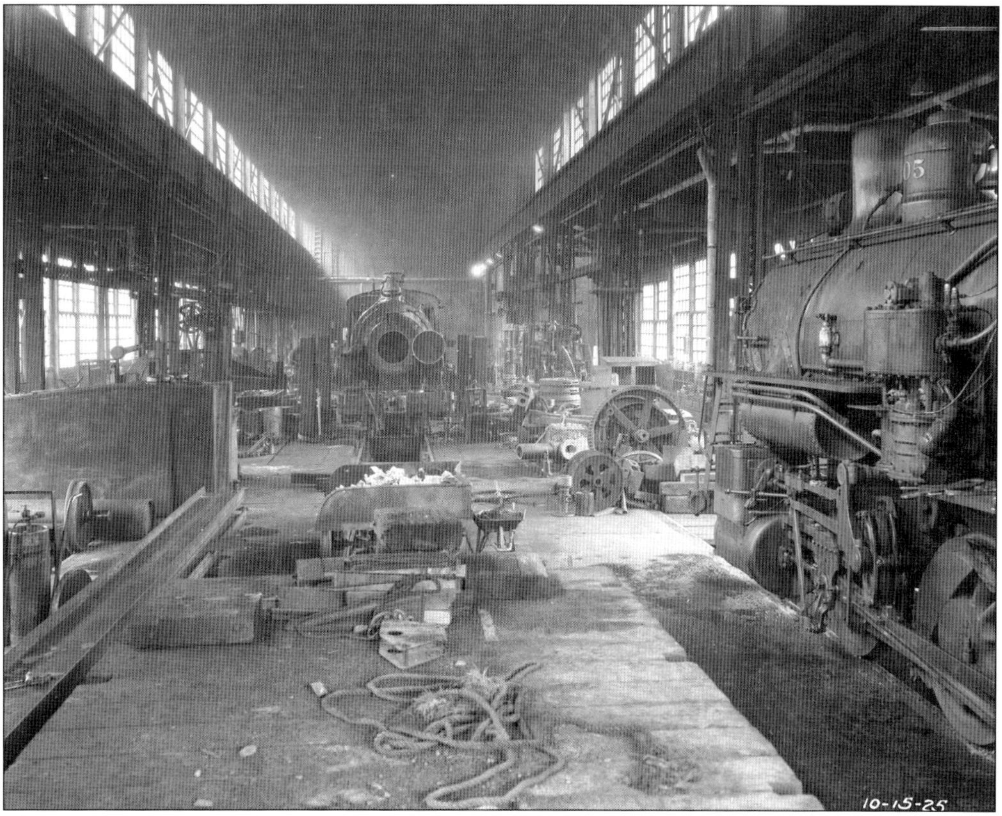

This is a rare look inside Utah Copper's machine shop. Utah Copper did all its own work, maintaining its roster of steam and, later, electric locomotives. This photograph is dated October 15, 1925. Soon, the open-cut mining operation's steam-powered shovels and trains were converted to electric. Only the main line trains to the mills would use steam then and do so until 1948, when the all-electric new Copperton Low-Line was put in service.

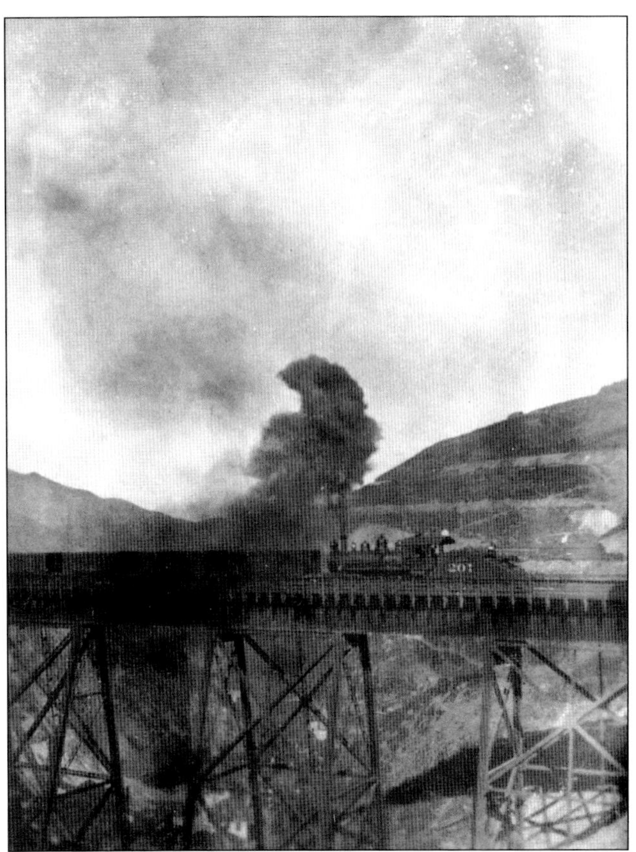

Bingham & Garfield's No. 201 crosses the Carr Fork Bridge, or A-line bridge. The No. 201 was a Baldwin 0-6-0 locomotive with a builder date of February 1907. The locomotive came to Utah Copper as No. 10 in 1907; it was moved over to the Bingham & Garfield Railroad in 1908 and renumbered to 201. The Bingham & Garfield Railroad was organized in 1908, and construction began in 1910; it was put into service in 1911.

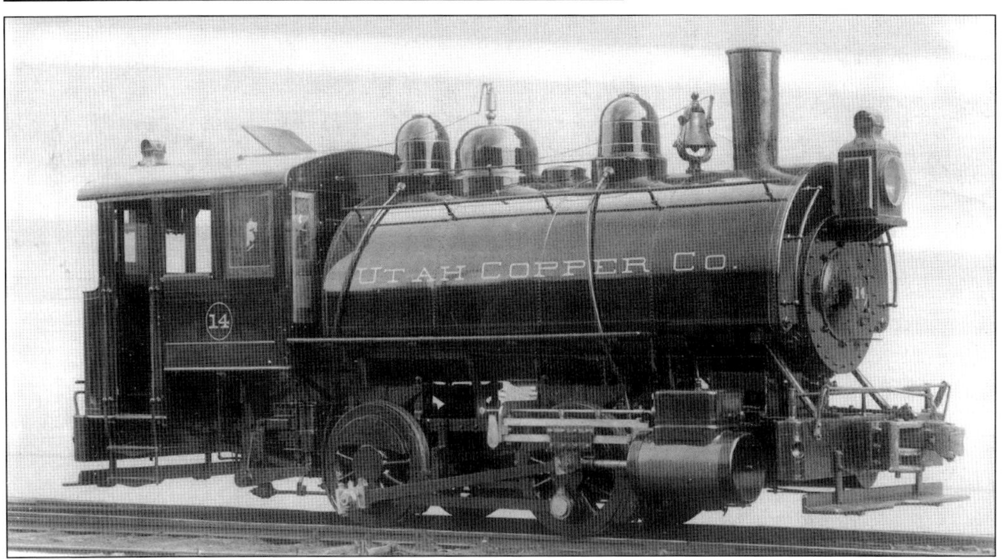

This is a close look at Utah Copper No. 14, a Porter 0-4-0 saddle tank, known as a 50-ton locomotive. These little steam locomotives dinkeys gave Utah Copper years of good service, but the company was always looking for a better and more economical way to move material. Steam engines had to have a constant supply of coal and water, and maintenance costs were high. In the 1920s, steam was on its way out.

This image shows the coal and ash towers in the B&G yard on November 11, 1930. By this time, electric locomotives are in use at the mine. Only the Bingham & Garfield Railroad uses steam locomotives to move the ore to the mills. The locomotives would dump their ash, which would fall through the tracks to the level below, and the ash would be carried up on a conveyor and loaded into a train car.

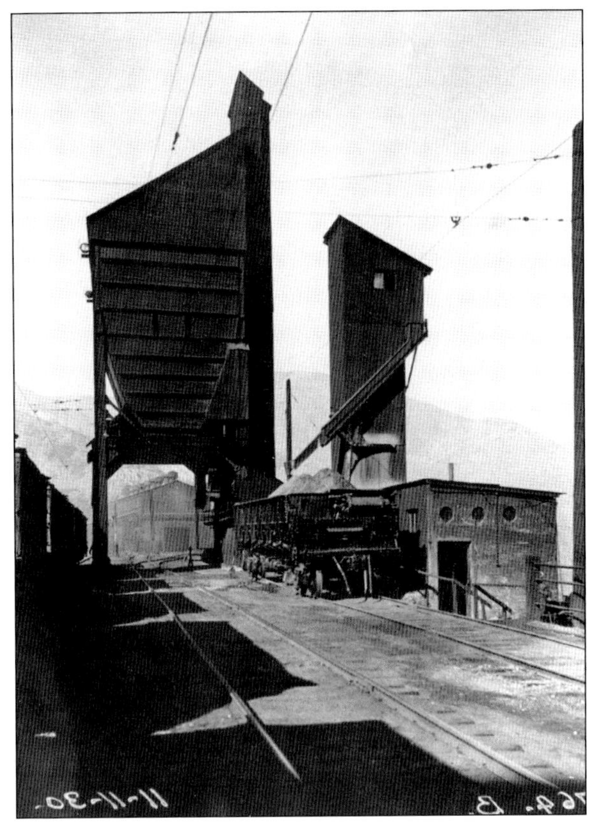

This view looks down on Utah Copper's staging yard around 1910. The dinkeys would bring a few loaded ore cars down from the mountain, drop them off, and take empty ore cars back to their shovel. The ore train would then cross over the S-type bridge (bottom right) to the Rio Grande Western Cuprum yard. More ore cars would be combined into a larger train for the trip to the mills.

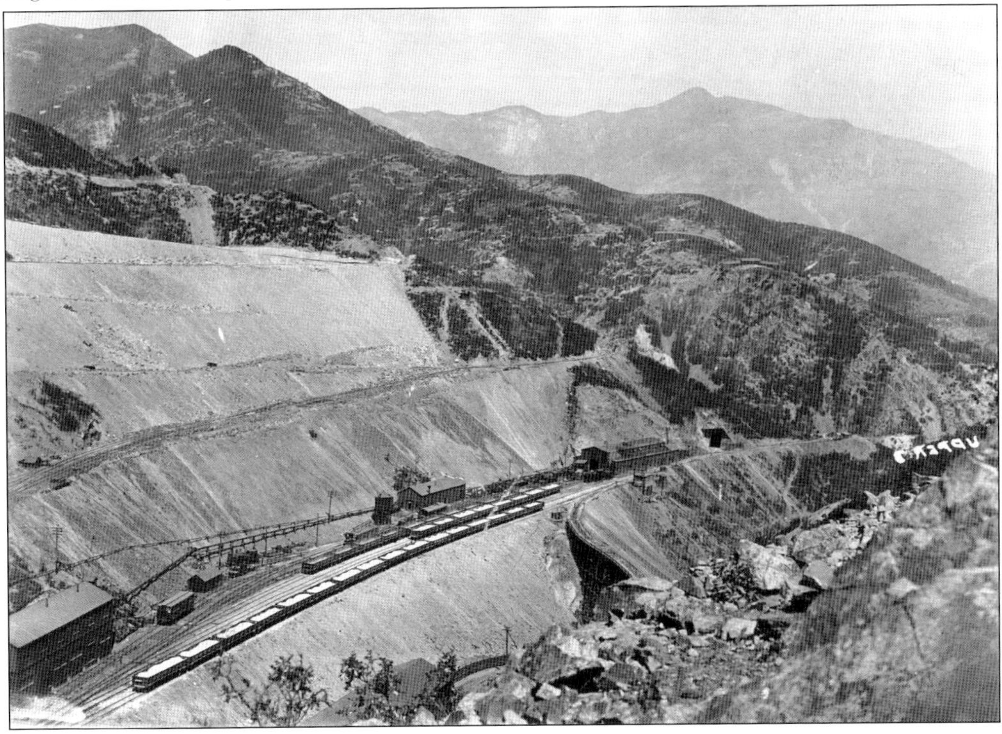

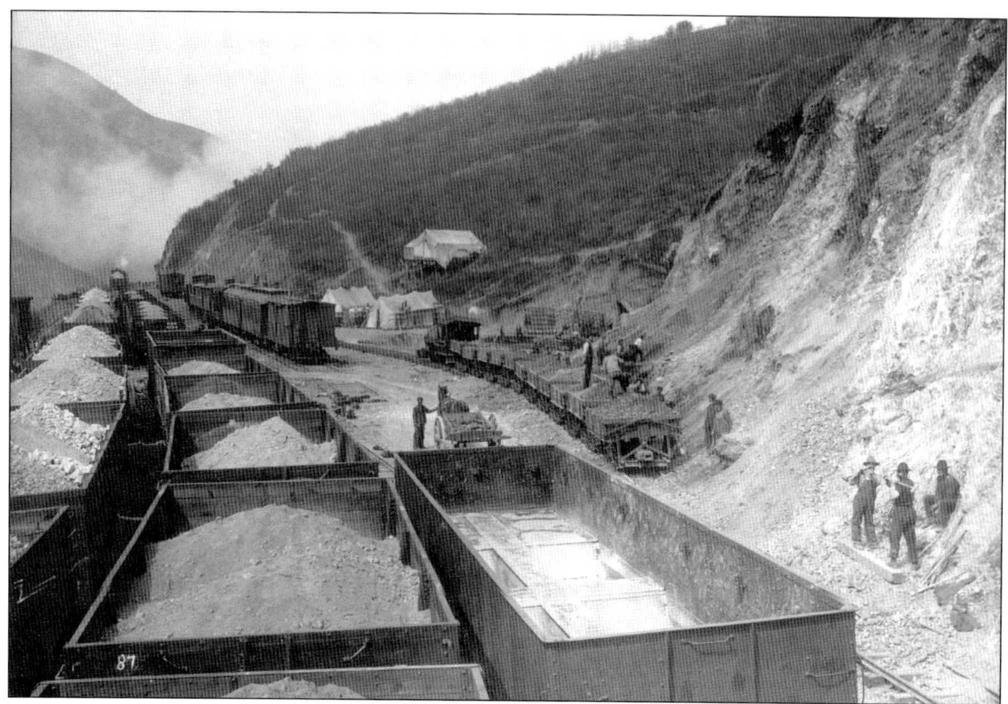

This image shows the busy Rio Grande Western Railway Cuprum yard. Both Utah Copper and Boston Consolidated were using the Rio Grande Western Railway to move their ore from the open-cut mining at Bingham to their mills at Magna. Daniel Jackling was dissatisfied with the Rio Grande Western Railway service, and this prompted Utah Copper to build its own railroad line, the Bingham & Garfield Railway. The B&G was organized in 1908, and construction began in April 1910.

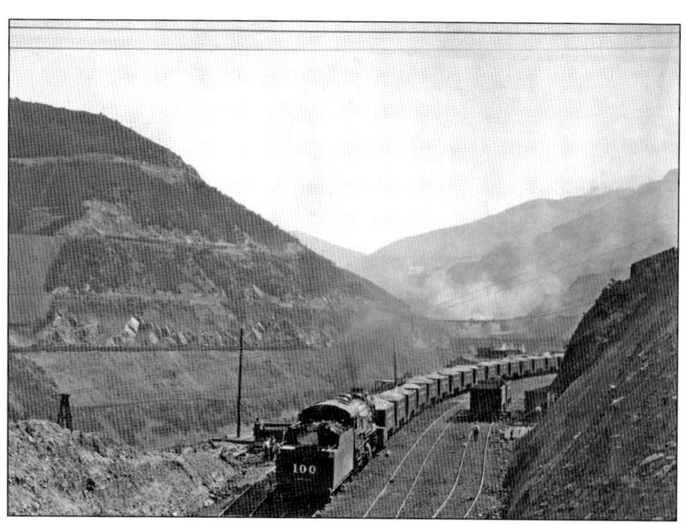

This historical photograph shows the first train on the new Bingham & Garfield Railway on September 14, 1911. This new line was approximately 20 miles long and cost $3,336,000 to build. Utah Copper built its own railroad line because the Rio Grande Western Railway could not keep up with the volume of ore coming out of the mine and refused to make improvements. Directly across the canyon is the Rio Grande Western line.

Bingham & Garfield Railway locomotive No. 101 is pulling a long string of ore cars across Markham Bridge in 1914. The impressive B&G took 15 months to complete. The line entered the canyon on the north side; once in the canyon, it crossed over the large Dry Fork Bridge, then went through a series of four tunnels totaling 4,795 feet, and then traveled over the Markham and Carr Fork bridges before entering the mine.

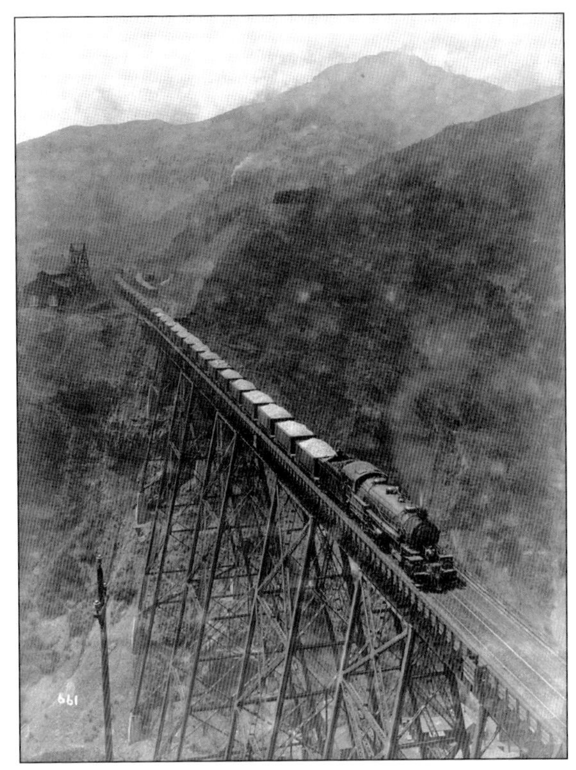

The photograph shows Bingham & Garfield locomotive No. 104. The B&G line used some of the largest locomotives for its time. No. 104 was an Alco-Schenectady 0-8-8-0 Mallet compound locomotive with a builder date of March 1917. The locomotive could manage 40 loaded ore cars, approximately 65 tons each, for the hour-and-a-half trip to the mills. Two train crews could deliver around 10,000 tons of ore a day.

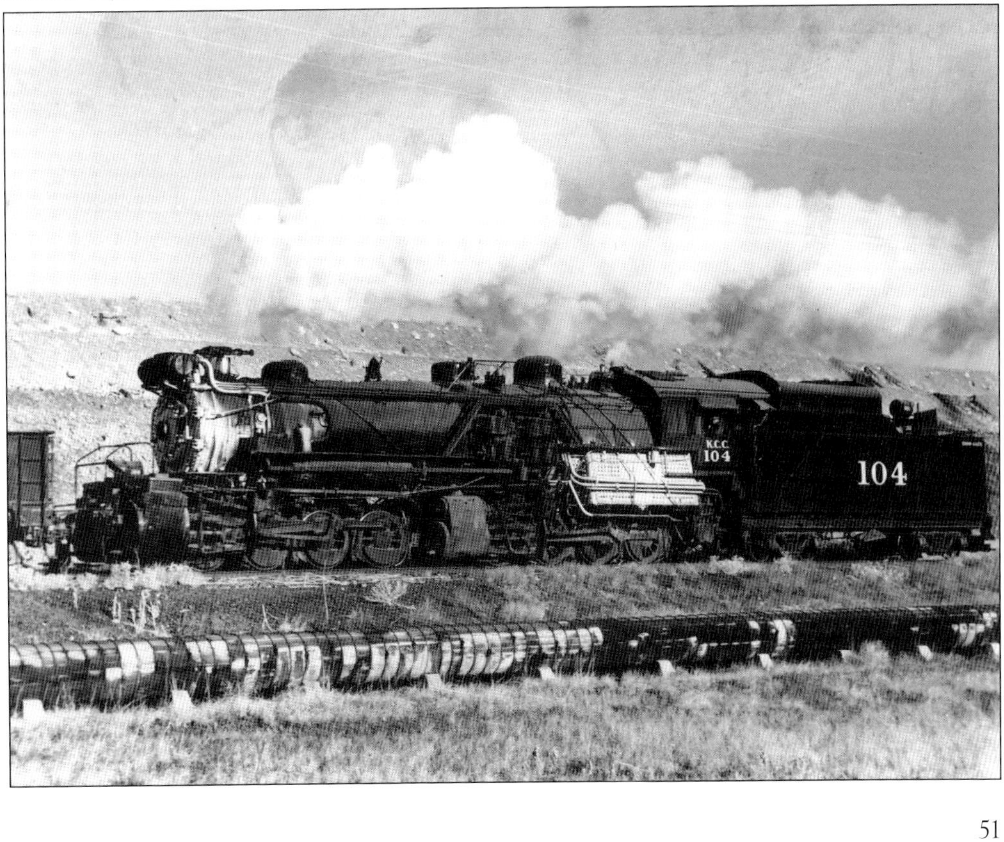

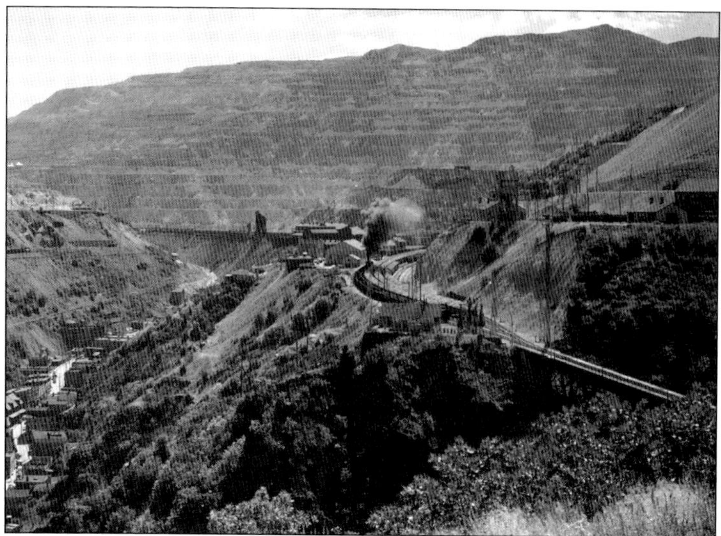

This is the view when entering the mine from the Bingham & Garfield Railway in 1938. The large coal and ash towers are in the center of the photograph. The Markham Bridge, 640 feet long and 220 feet high, is pictured at bottom right. The staging yard for the B&G has now moved to the north side of the Carr Fork Bridge. The town of Bingham is on the canyon floor below.

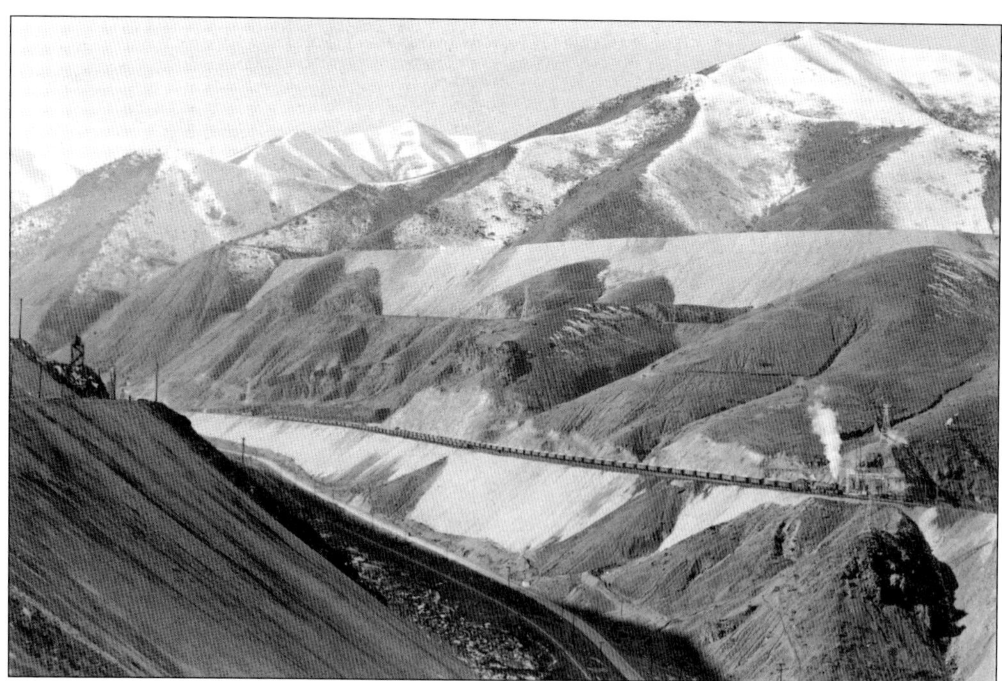

This image shows a long ore train traveling along the newly constructed cross-canyon connection in 1945. The train would meet up with the old Bingham & Garfield line shortly. Once open-cut mining went below A-level, a series of tunnels in the mine started to be built to keep the ore moving downhill. The cross-canyon connection was constructed to connect the new tunnels with the old B&G line.

Five

Electric Trains and Shovels

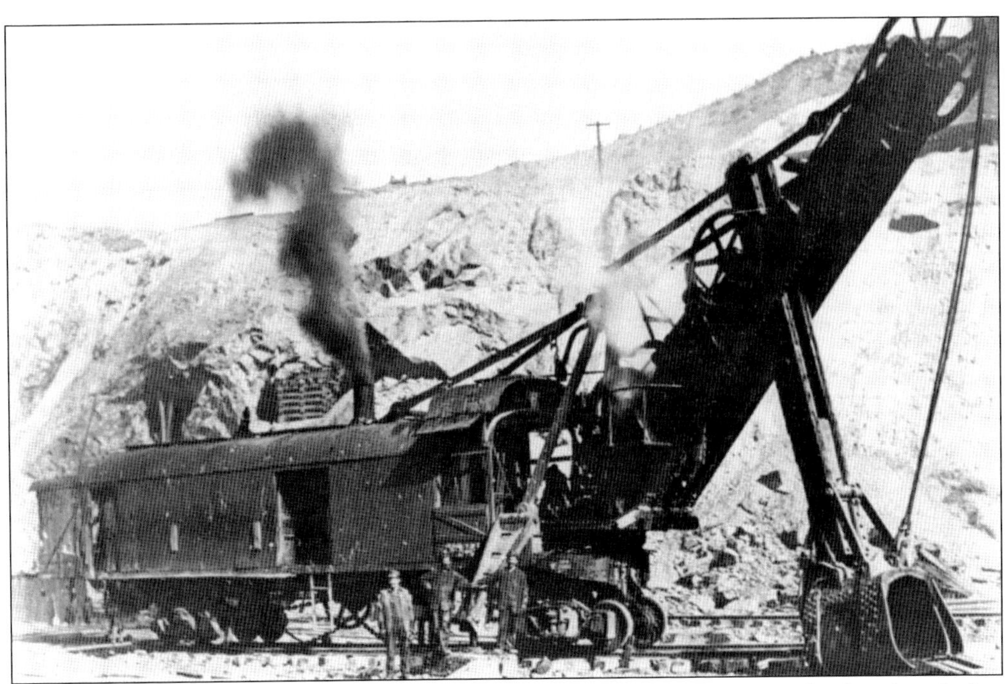

The image shows a standard railroad-type shovel at Bingham Canyon Mine. The first shovels that came to the mine in 1906 operated on steam and ran on rails parallel to the trains. A report by Andrew Hodges in 1925 said that it took four men to lay rail in front of the shovel and a six-man crew to run the shovel—that is a total of 10 people to keep the shovel moving ahead.

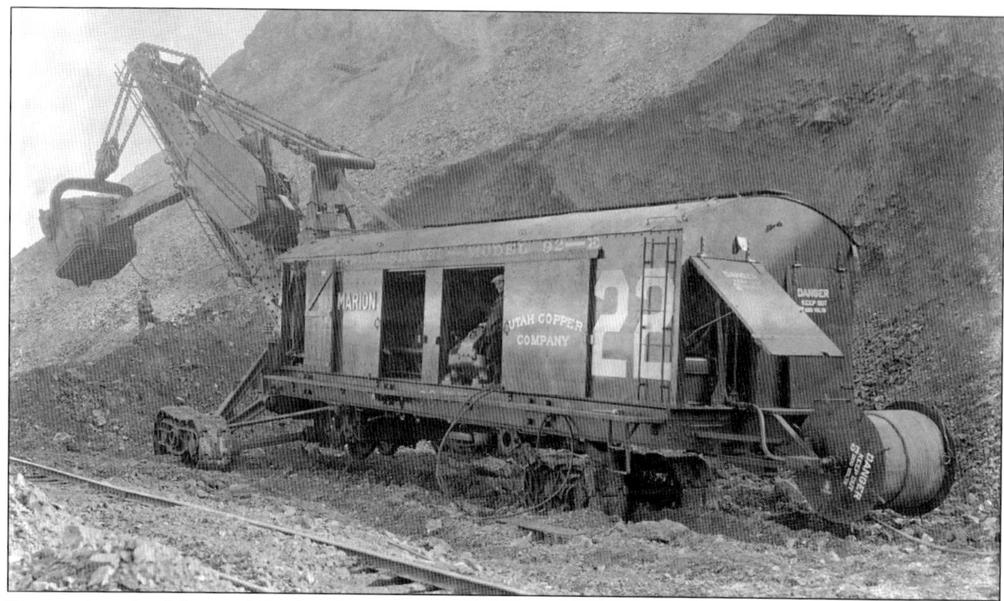

Pictured is the Marion No. 22 electric shovel at Bingham. Note the cable reel on the back. The first shovel to be converted from steam to electrical power was in 1922, and the first electric shovel purchased was in 1923. Both were railroad-type shovels but were mounted on crawler treads, with 4.5-yard dippers. The saving was 10¢ per ton, which prompted the decision to convert the entire roster of shovels to electricity.

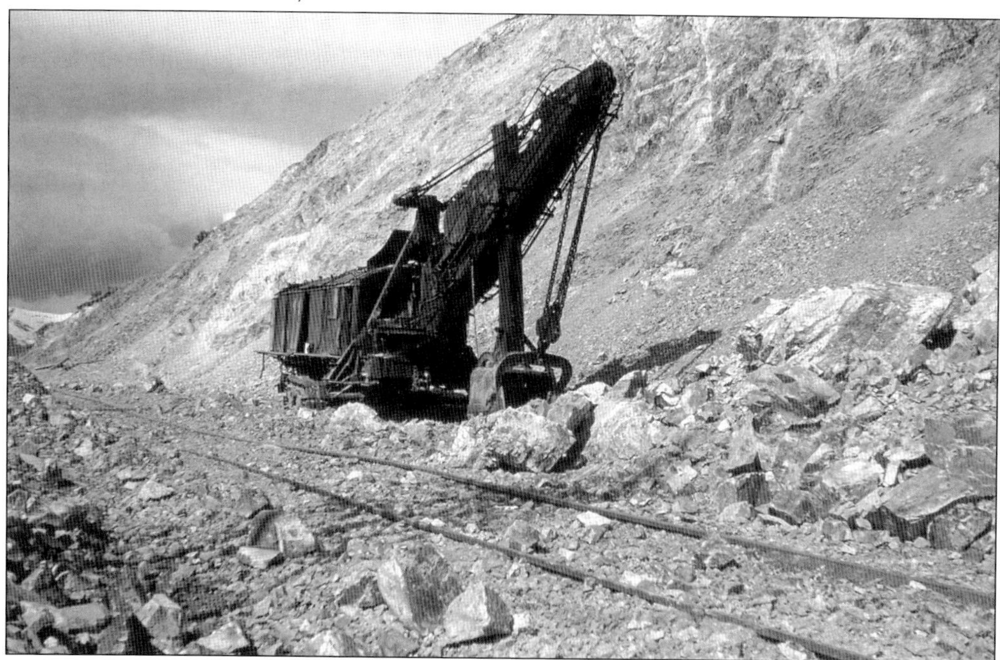

Pictured is an old standard type of shovel, still in use in March 1954. The first two electric shovels at Bingham were model 92 Marions, delivered in 1923. One of the shovels had alternating current, and the other one had direct-current motors. After a year of testing, eight alternating-current shovels were ordered, and eight steam shovels were converted to electric power. By 1927, there were 23 electric shovels in operation, with no steam shovels.

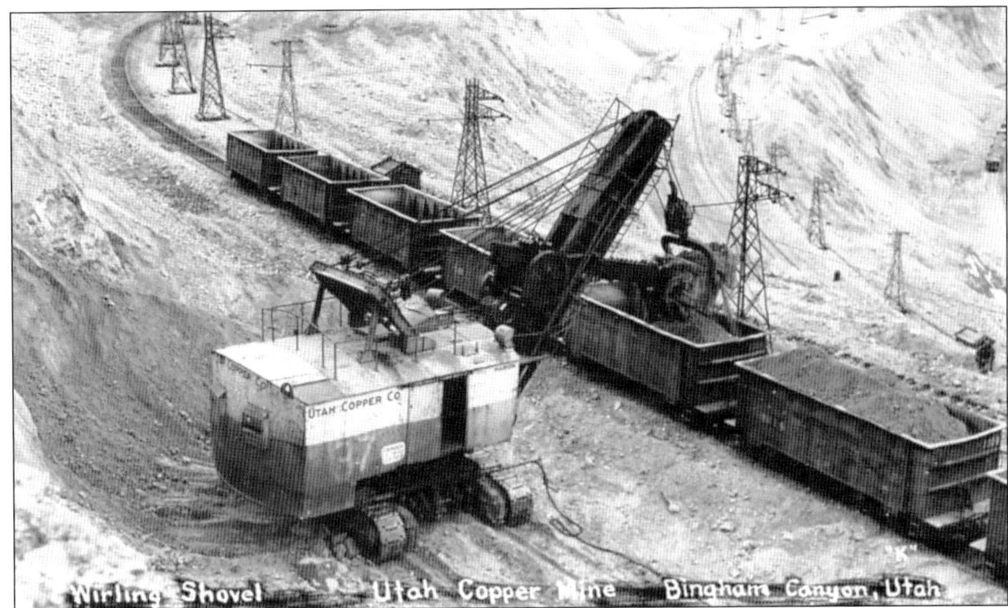

This image shows a Marion full-rotation shovel, nicknamed a "whirling" shovel (spelled here as "wirling"). In 1927, the first full-rotation Marion model 4160 rope shovel came to the mine. Marion introduced model 4161 in 1935, and over the years, Marion kept making improvements, increasing capacities with model 151-M in 1945 and model 191-M in 1951. On January 29, 1939, Utah Copper reported having 29 electric shovels; six were full-rotation type, all equipped with 4.5-yard dippers.

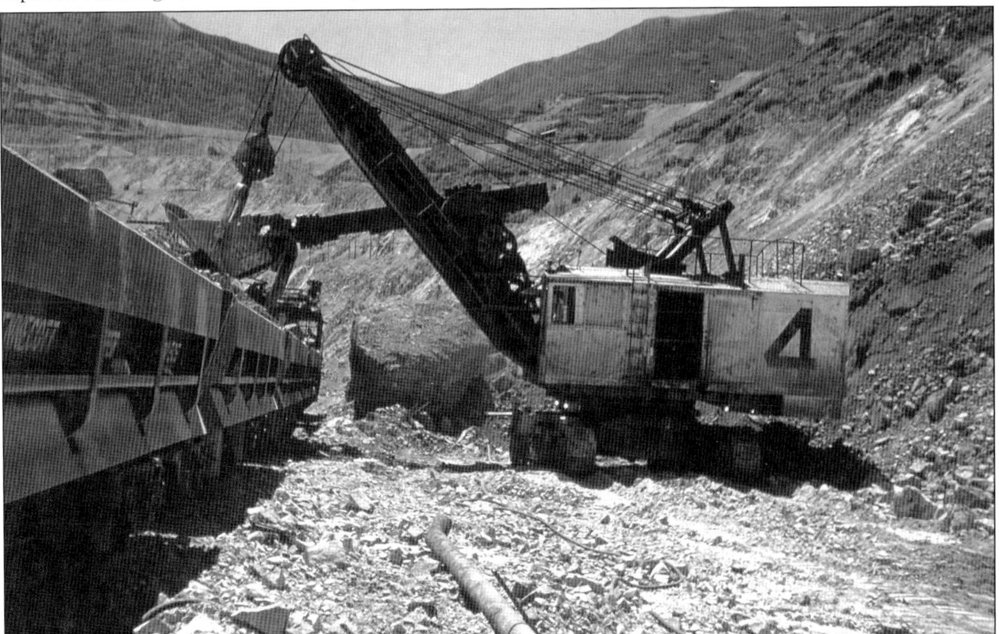

The No. 4 electric shovel loading waste cars are pictured here in September 1956. By late 1940, Utah Copper's daily ore production was 70,000 tons, with 90,000 tons of waste rock moved. Electric shovels were loading 6,300 tons each eight-hour shift. The new full-rotation shovels with five-cubic-yard dippers loaded 10,000 tons per shift. The pit shovels remained small so as not to interfere with the electric locomotive catenary line. The last standard shovel retired in 1954.

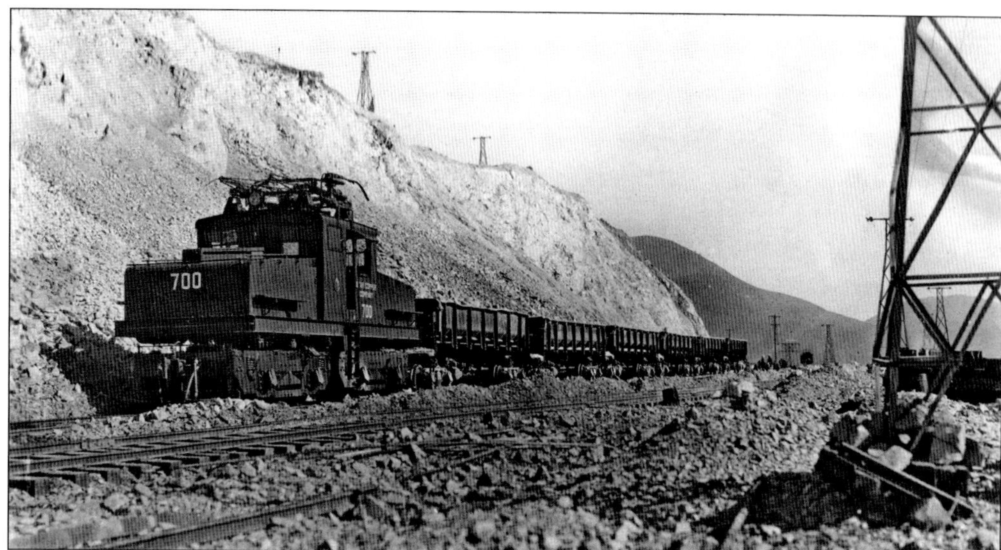

The battery-operated electric locomotive No. 700 is away from the catenary line. On May 16, 1927, the No. 700 electric locomotive went to work on the K-level dump line. The work was part of motive power tests with steam, diesel, oil-electric, and electric motors at the mine. The General Electric (GE) 750-volt DC locomotive won out. The Utah Copper, later the Kennecott Copper rail system, became the largest privately owned electric railroad in the world.

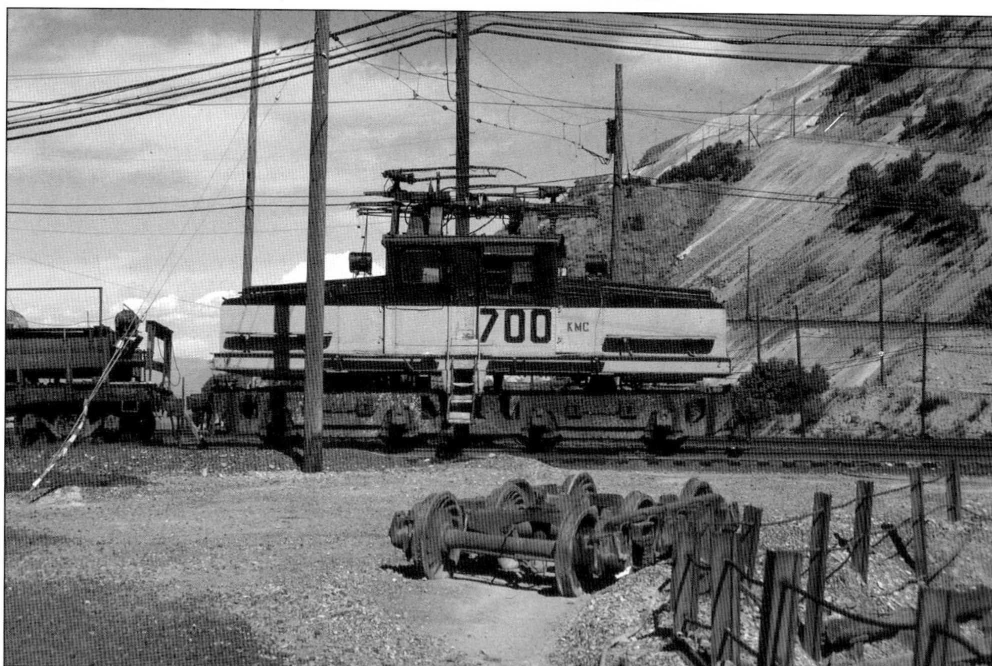

This image shows electric locomotive No. 700 in 1984, with a builder date of May 1927. By the time of the photograph, it had been around for a long time. The roster of electric locomotives gave Utah Copper many years of great service. Once Utah Copper made the decision to use electric locomotives, from 1927 until 1929, they ordered 42 General Electric 85-ton, 750-volt DC locomotives. Over the next years, more groups of electric locomotives were ordered. The last four came to the mine in 1955.

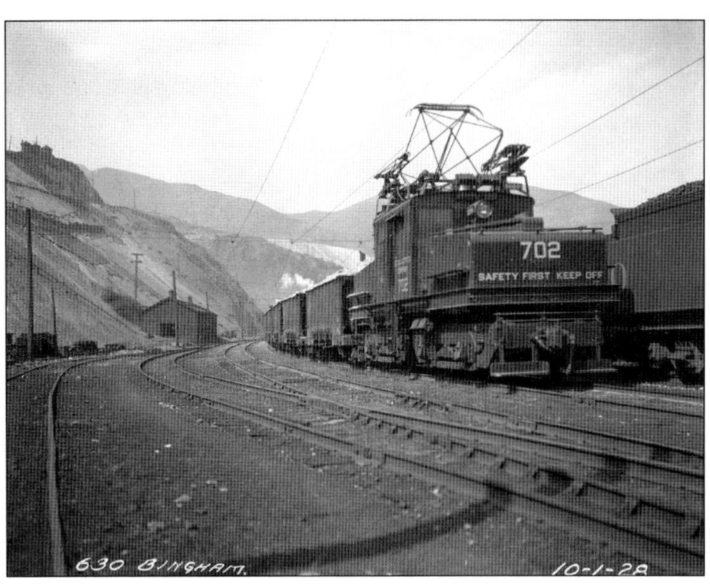

This image shows the No. 702 electric locomotive in the Bingham & Garfield Railway yard on October 1, 1928. The builder date is September 1928; in that year the electric locomotives had an all-black paint scheme. This photograph shows the overhead pantograph in use; the two side-arm pantographs are folded down next to the cab. The overhead pantograph extended 17 feet, 7 inches to 26 feet. The side-arm pantographs extended 17 feet, 7 inches to 22 feet.

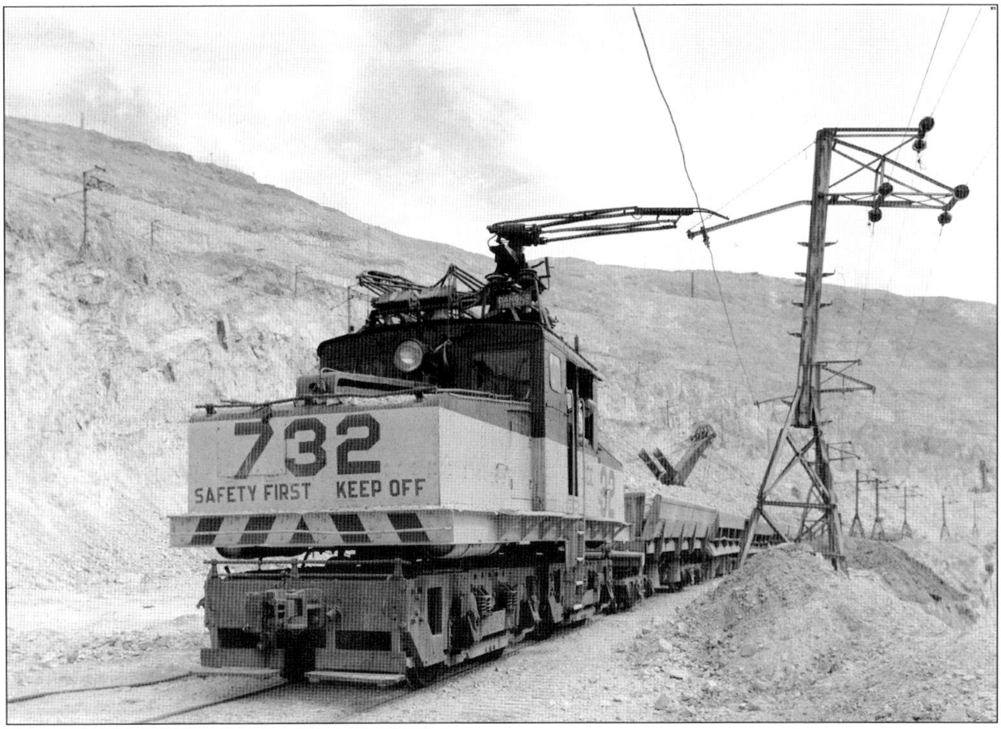

The No. 732 electric locomotive and waste train are pictured at the shovel. This image shows the side-arm pantograph extended; this was important so as not to interfere with the shovel dipper. The black-and-yellow paint scheme on the locomotive and the type of electrical towers in use mean it is the 1960s or 1970s. Note the large 5,000-volt flexible armored trail cable on the ground; it supplied electricity to the shovel, bottom left.

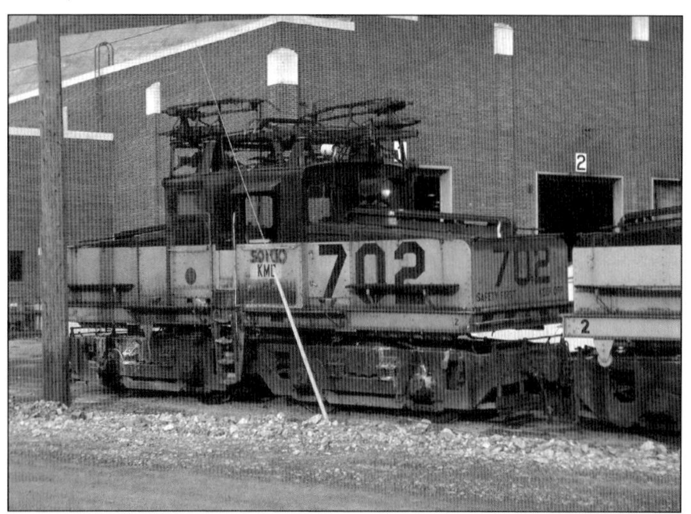

The No. 702 steeple cab electric locomotive sits outside the Dry Fork train shop. The train shops moved down to Dry Fork Canyon in 1947. The old Utah Copper train shop that was perched on the mountainside had to go, as the mine expansion was moving its way. The machinists and mechanics kept the large roster of electric locomotives on the rails, repairing everything down to rewinding the electric motors.

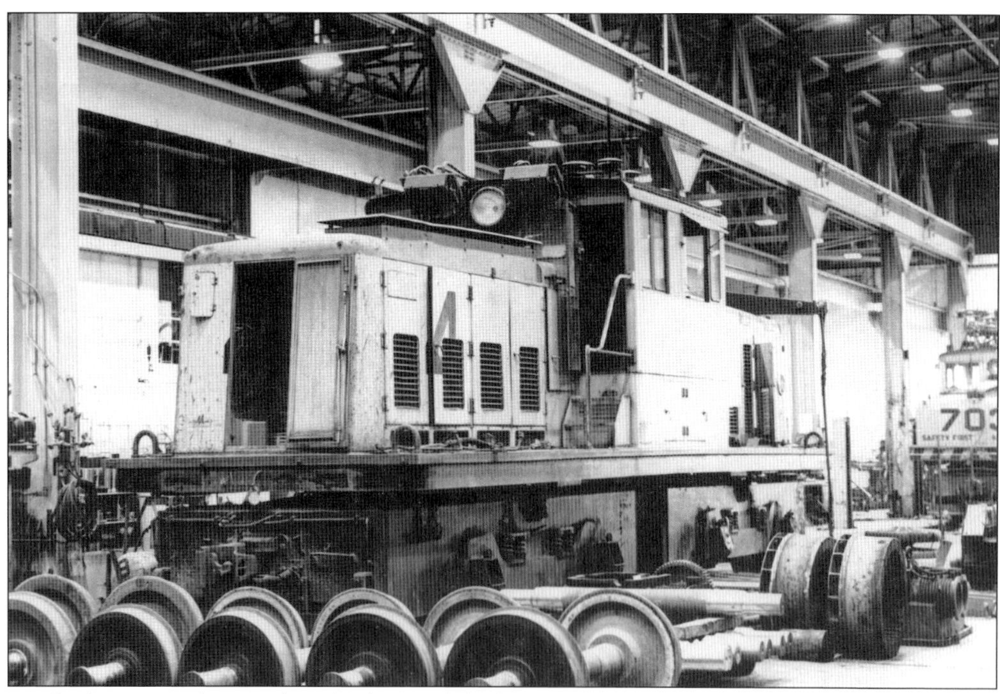

Inside the Dry Fork train shop (nicknamed the "New Shops"), machinists rebuild the Kennecott Chino Mine's No. 4 electric locomotive. It was transferred to Kennecott Utah in December 1971, where it was renumbered to 778. In May 1976, it was painted in a red, white, and blue paint scheme to commemorate America's bicentennial. Then repainted back to standard yellow and black in June 1978. The 778 was a 125-ton locomotive with a builder date of April 1958.

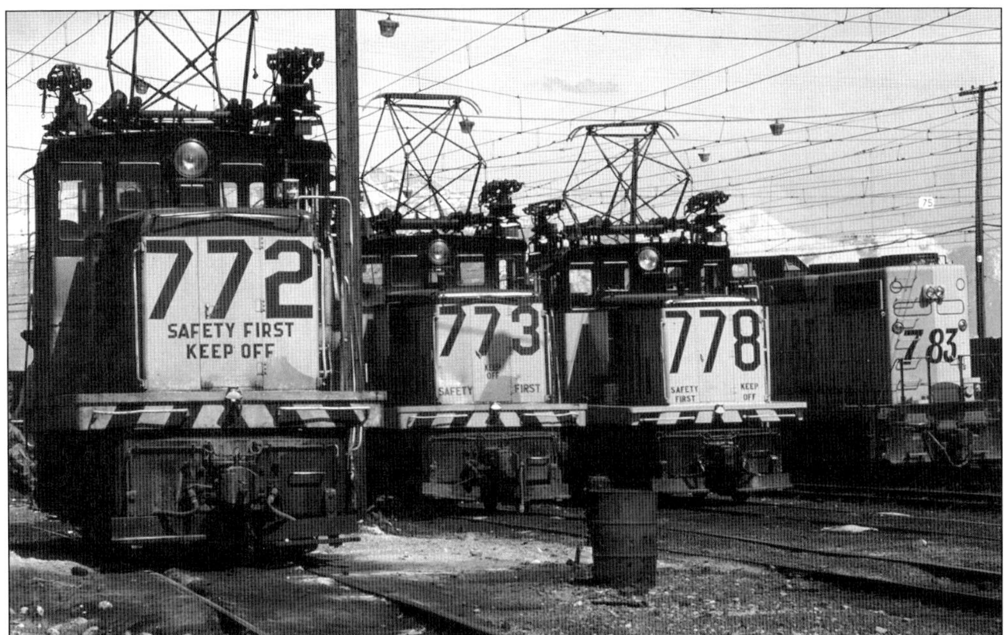

Three electric 125-ton locomotives are sitting in Copperton yard in May 1978. Over the years, more groups of electric locomotives came to the mine. In 1937, the mine received a group of nineteen GE 85-ton electric locomotives; in 1942, four GE 90-ton electric locomotives; in 1952, four GE 125-ton electric locomotives; and in 1955, four GE 125-ton electric locomotives. Note the fire barrow in front of the locomotives for those cold nights. (James Belmont.)

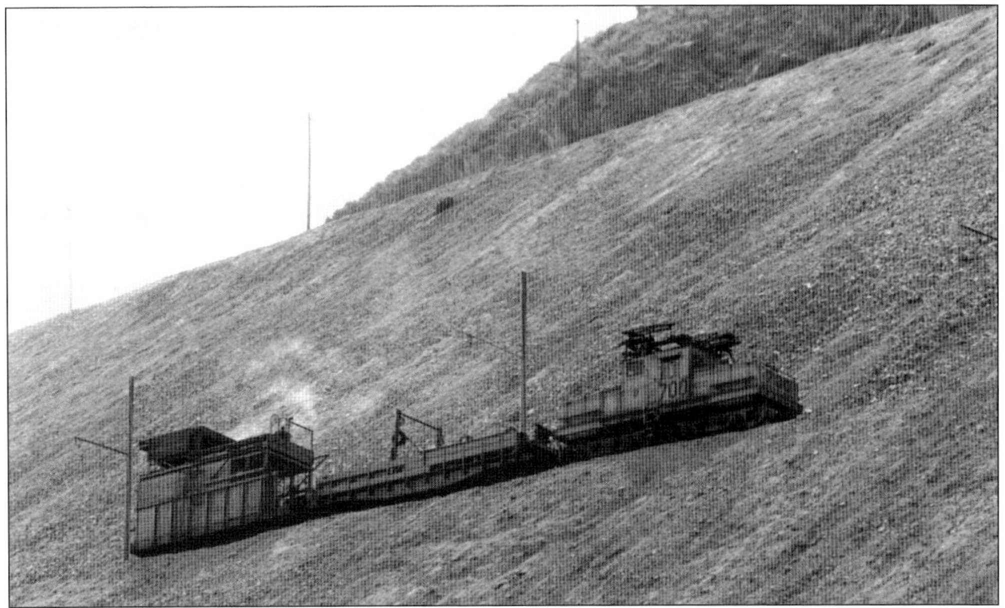

This image shows the No. 700 electric locomotive at work with the line car. The locomotive could work away from the catenary line because it was also battery-operated. The other locomotives with batteries were Nos. 702, 703, and 724; they had a 500-hour charge and could recharge through the catenary line. The battery-operated locomotives could work away from the catenary line when working with cranes, the tool car, and the line car.

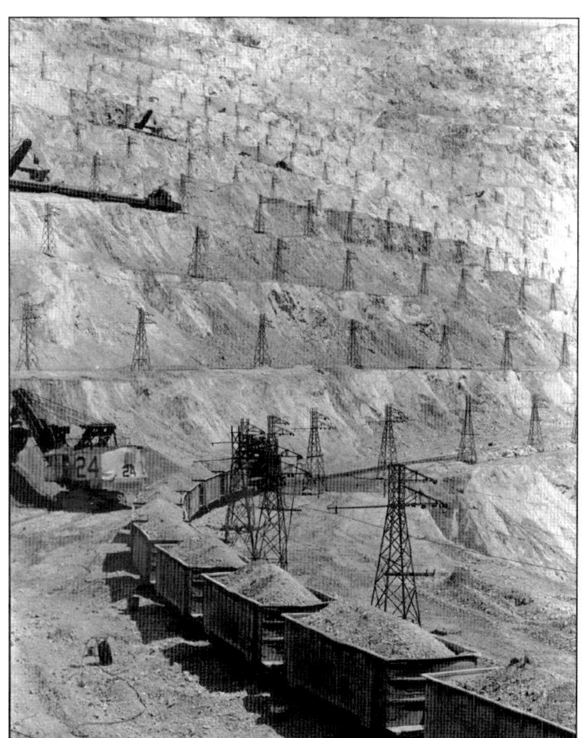

A loaded ore train pulls away from the No. 24 shovel in the open-pit operation in the mine. This was an everyday occurrence, 24 hours a day, seven days a week. Ron Burke, a train foreman, said that, at one time, there were 22 train crews per shift. In 1940, the mine had 120 miles of track. Most of that trackage was temporary and had to be moved constantly.

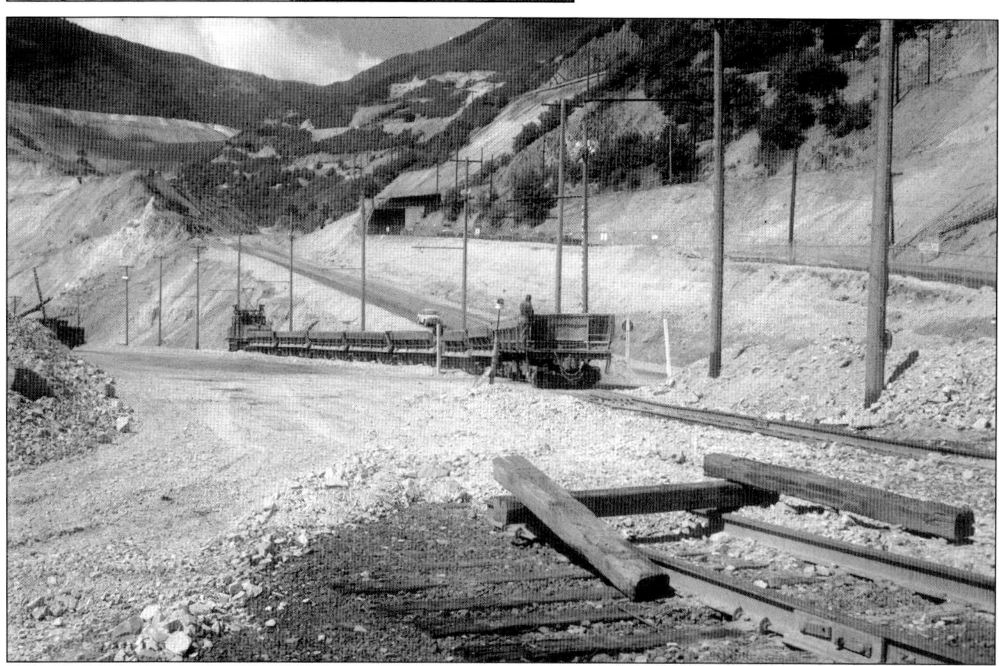

This image shows an empty waste train pulling into a level. The train is on the north side of the mine at Carr Fork Canyon, once home of Highland Boy. Note the haulage truck center left; it indicates the year this image was captured as after 1963, when truck haulage came to the mine. The waste train has eight waste cars. These trains were usually shorter than an ore train, which would have anywhere from 10 to 20 ore cars.

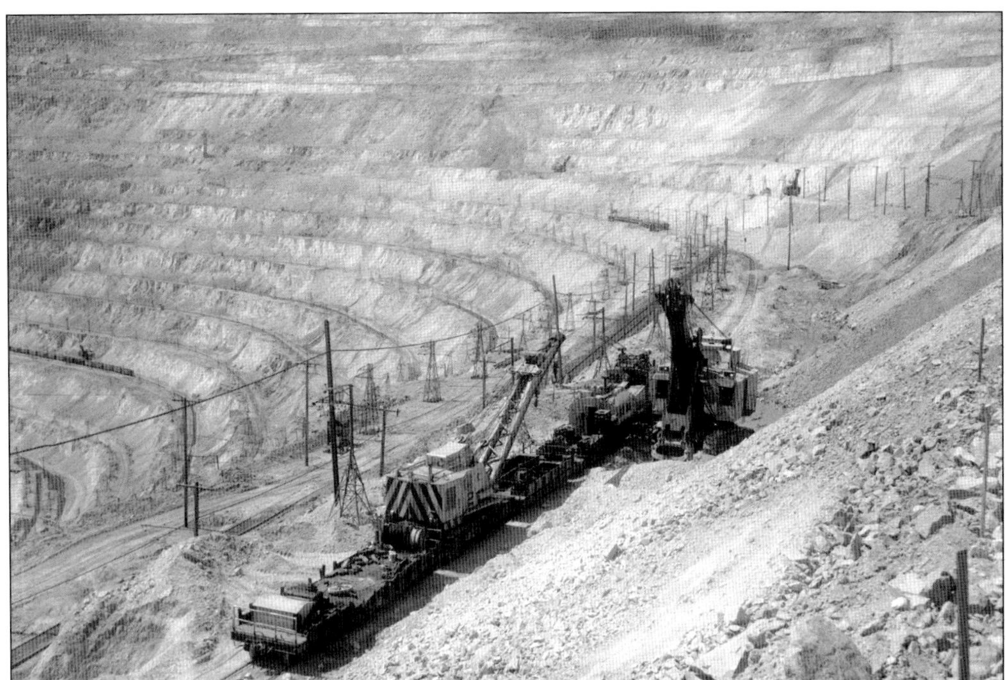

In this June 11, 1965, photograph, the bull gang repairs a shovel somewhere in the mine. The bull gang did all the heavy work at the mine, building and repairing the electric shovels, and were first on the scene when there was a wreck. The gang would travel around the mine with the derrick crane (the big hook) and tool car, as shown here. It was imperative to keep the shovels working.

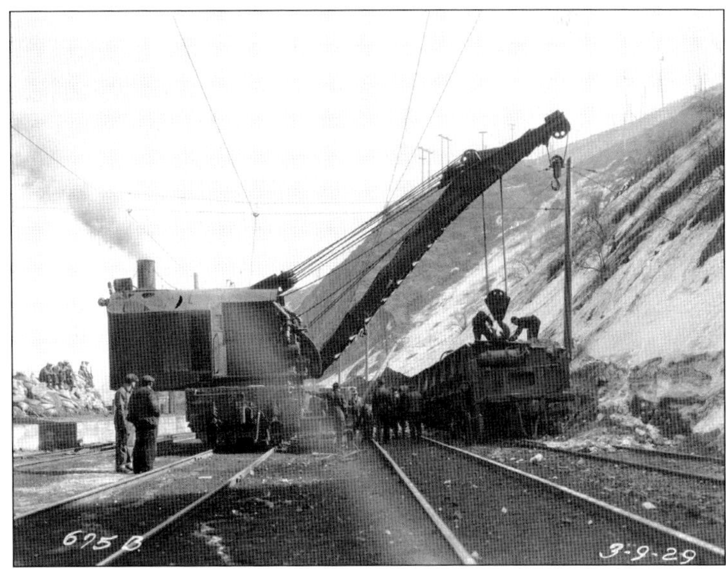

This is an early photograph of the bull gang at work. These men hook up cables around the waste car so that the derrick crane can put the car back on the tracks. The date is March 9, 1929, and the big hook (the crane) is steam-powered. Kennecott mechanics would convert their derrick cranes to diesel power. The men sitting on the berm to the left are probably the track gang, waiting to repair the track.

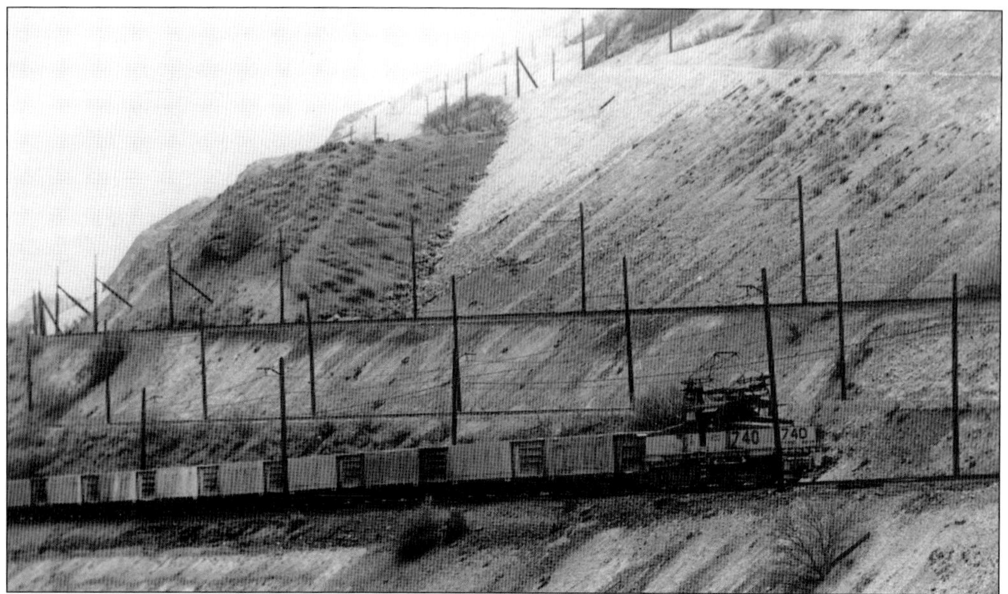

The No. 740 electric locomotive grinds its string of empty ore cars up Bingham Canyon on the "four percent," the nickname given to this railroad line. After 1948, ore trains coming in and out of the mine went to the Copperton train yard. This is where the new main line to the mills started, called the Copperton Low-Line. Bringing a 2,700-ton train full of rock down the four percent took skill.

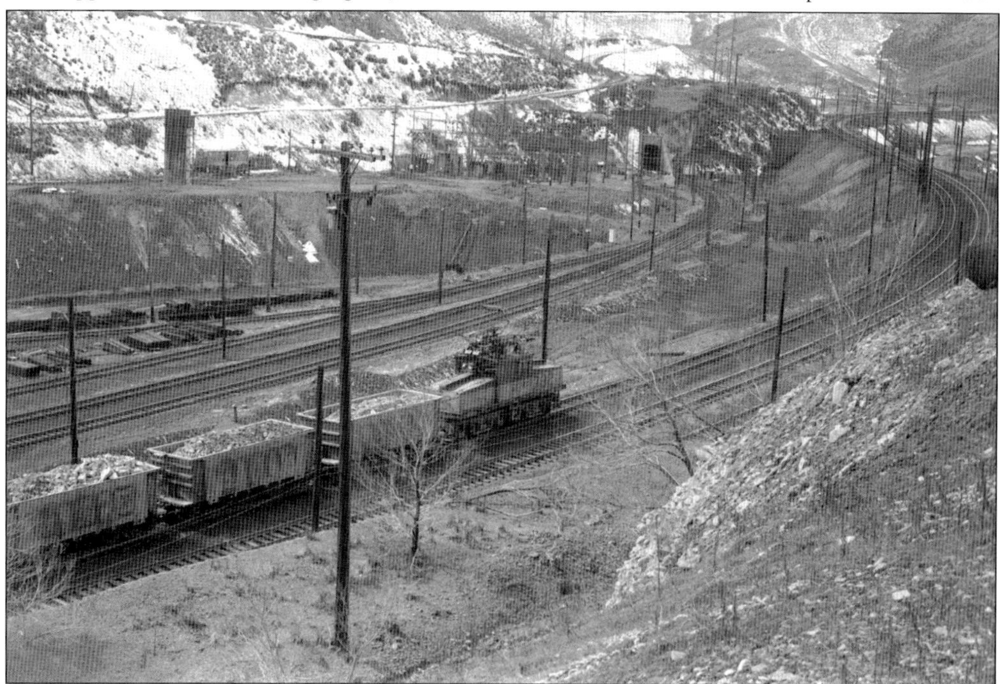

Electric locomotive No. 732 is entering Copperton yard, having just come down the Four Percent. This photograph shows the exit portal of the 5490 train tunnel that was in the mine at top right. The 5490 train tunnel was 18,000 feet long, making it the longest train tunnel in Utah. By this time, there were three tunnels built in the pit—the 6040, 5840, and 5490—all constructed to keep the ore moving downhill.

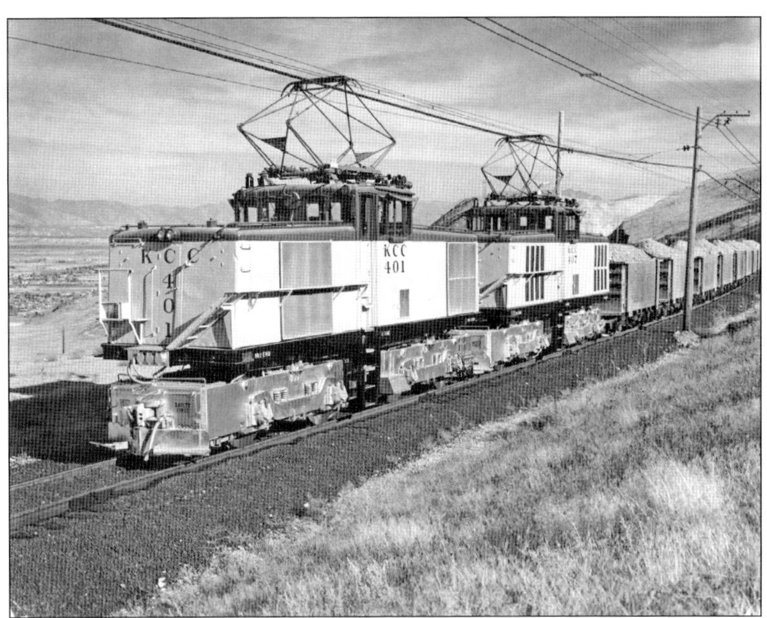

The Nos. 401 and 407 General Electric 125-ton locomotives pull a long string of ore cars to the mills. After trains of 18 to 20 ore cars brought their cars down to Copperton yard, they were made up into a larger train of 70 to 80 cars for the trip to the mills. The new Copperton Low-Line started operations on May 2, 1948, replacing the old Bingham & Garfield line.

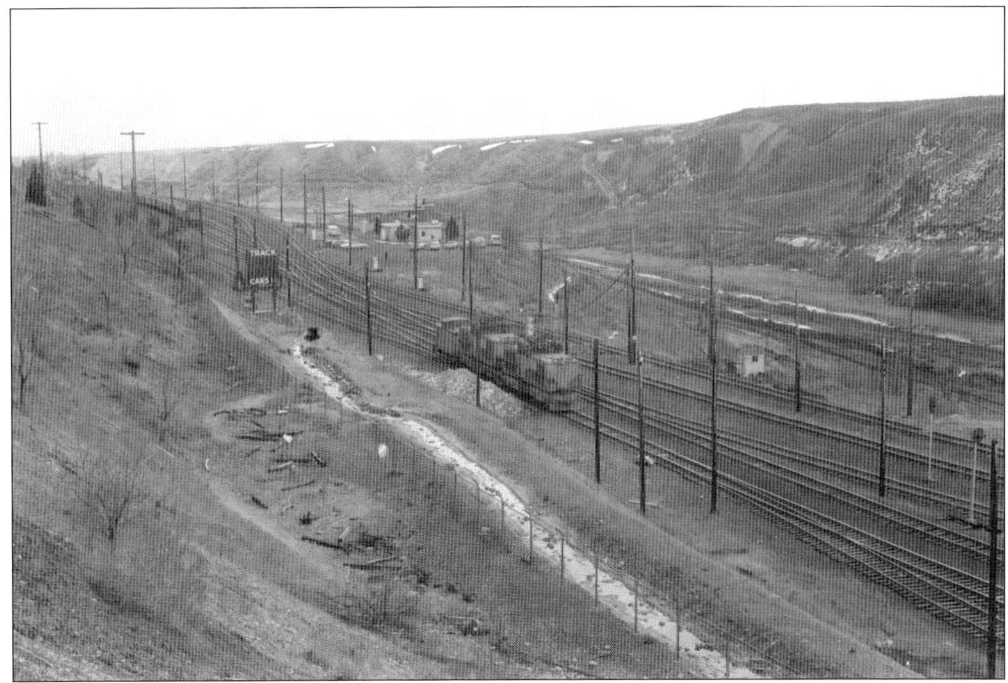

This image shows two class 400 General Electric 125-ton, 3,000-volt DC, 3,200-horsepower locomotives with a caboose in Copperton yard. Ore haulage always ran as a two-unit set and always had a caboose, which was cut off on the fly. Seven of these units (Nos. 401–407) came to the mine in late 1947. Working around the clock on this 28-mile round-trip, 1.35-percent downhill grade line, trains made 18 trips a day.

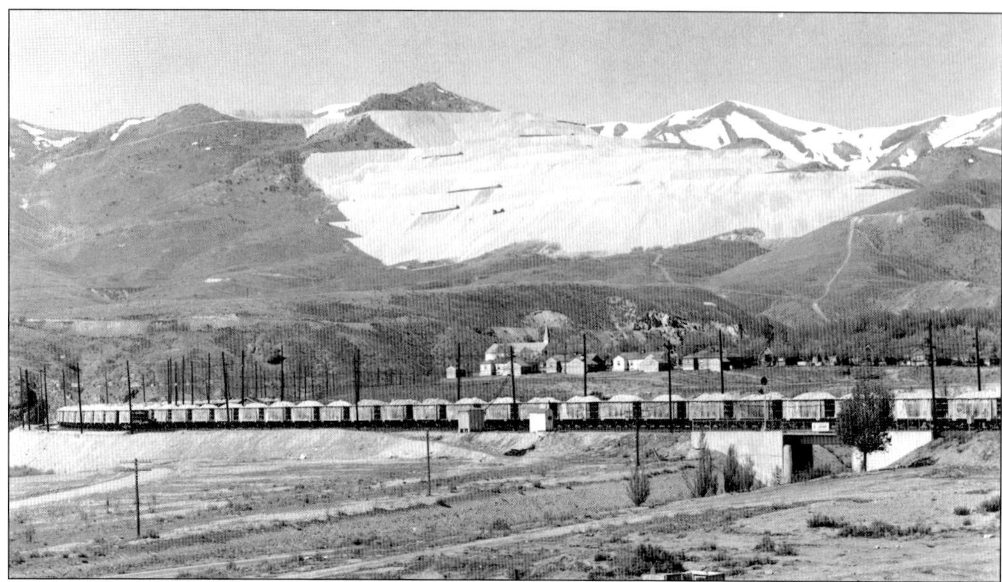

The ore train is leaving Copperton yard, which was crescent-shaped and wrapped around the small company-owned community of Copperton. The overpass was the entrance to Copperton and Bingham Canyon to the mine. The overpass was where the large Magna motors would energize, picking up 3,000 volts on the catenary; the two electric locomotives could be heard ramping up. The Copperton yard catenary was a lower 750 voltage to accommodate the pit locomotives.

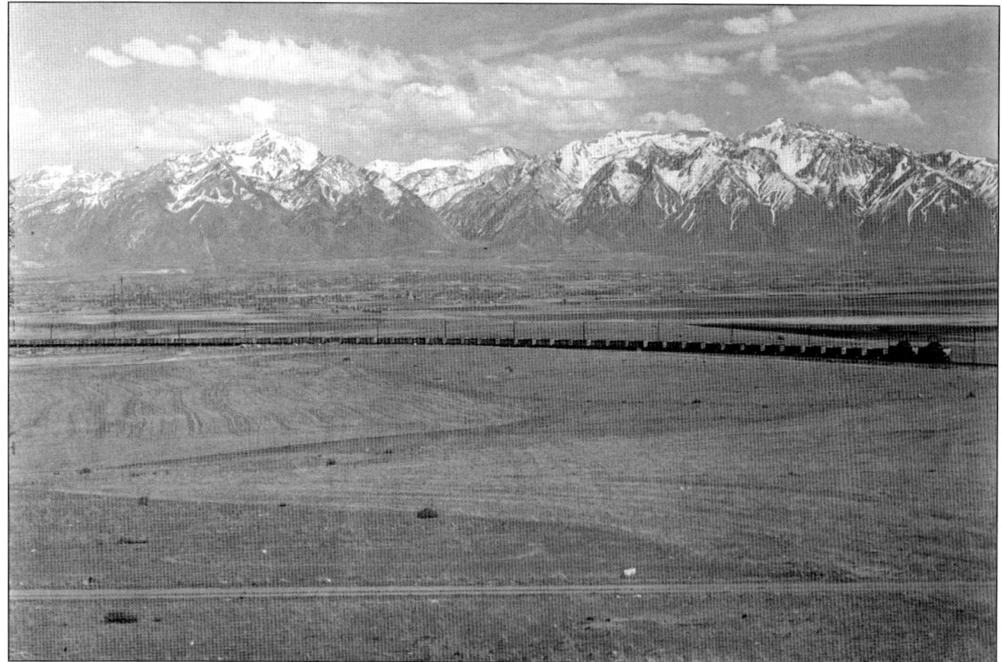

This image shows a long ore train and the Wasatch Mountains in the background. The empty ore cars are heading to Copperton yard, where they will be dropped off, divided up, and taken back to the mine. The Copperton Low-Line transported an average of 1,200 eighty-five-ton ore cars a day; that is 102,000 tons daily over these rails, making this line one of the busiest and the heaviest gross-tonnage-hauled railroads.

Six

Bridges and Tunnels

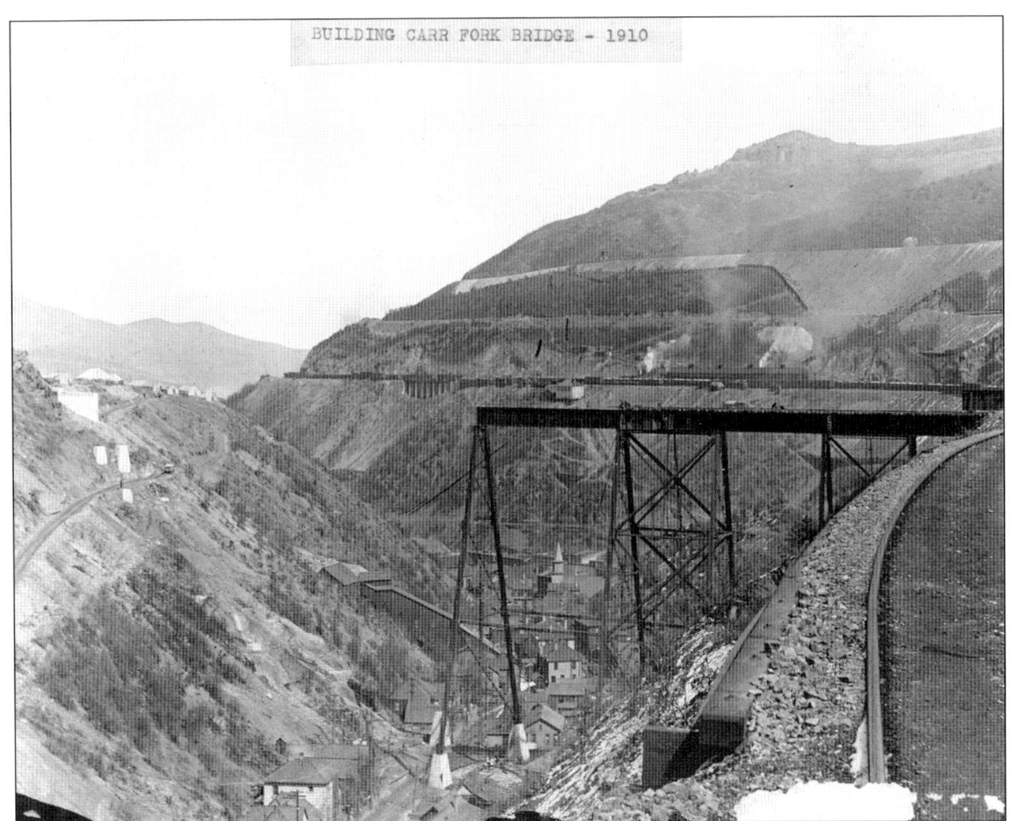

The image shows construction on the Carr Fork Bridge, also called the "A-line bridge." A-level was the bottom elevation at the mine. The early open-cut mining had to attack a mountain with steam shovels and trains to get to the ore and then bring the ore down the mountain to A-level, where large trains were made up for the trip to the mills. To accomplish this, tunnels and bridges were built.

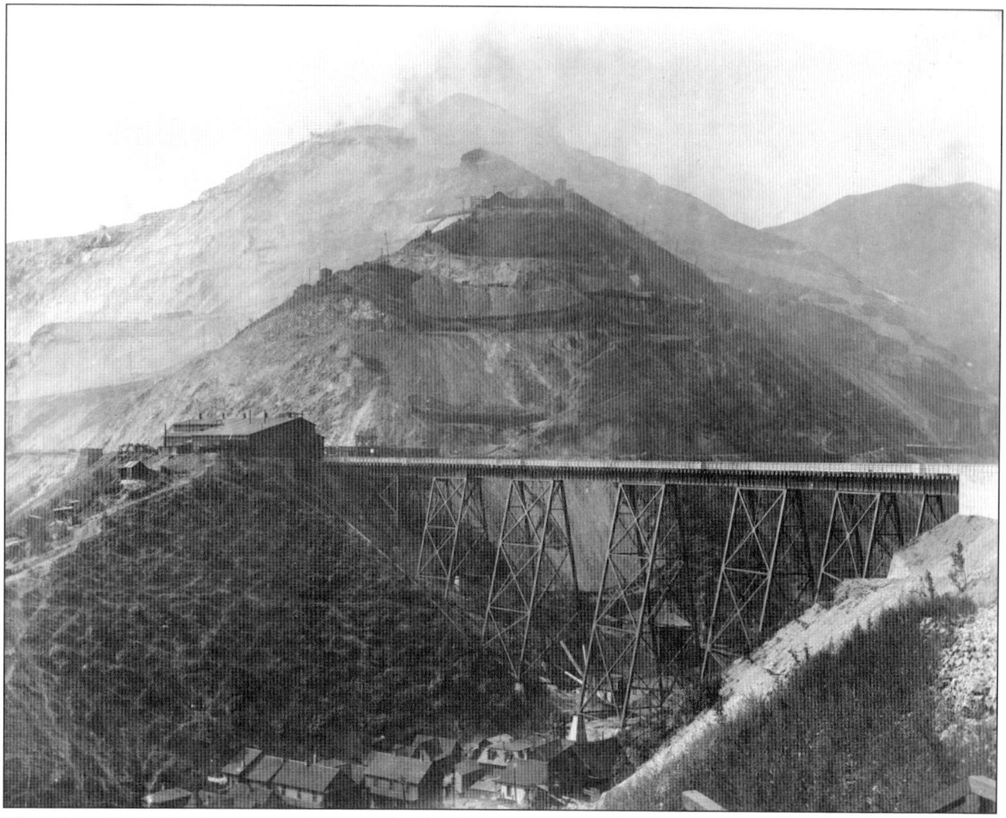

The Carr Fork Bridge is seen here in 1914. This was taken during the steam era, as evidenced by the amount of smoke on the mountain. The Carr Fork Bridge was the last bridge trains crossed before entering the mine. It was part of the Bingham & Garfield Railroad, the Utah Copper–owned main line to the mills. Carr Fork Bridge was a curved steel viaduct 680 feet long and 225 feet high, completed in June 1911.

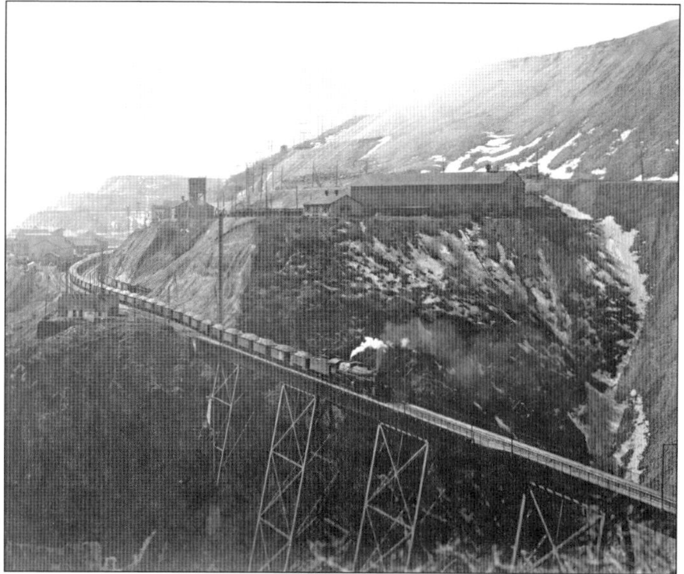

The ore train crosses over Markham Bridge after leaving the copper mine for its trip to the mill. Markham Bridge was the second of three large bridges built for the Bingham & Garfield Railroad. The name "Markham" came from the mountaintop, Markham Peak, and the bridge crossed over Markham Gulch. The Markham Bridge was a straight steel bridge 640 feet long and 220 feet high, completed in June 1911.

Ore trains crossed over Markham Bridge and immediately entered tunnel No. 4. This tunnel was 1,280 feet long and connected the Markham and Freeman Gulches. B&G tunnel No. 3 was straight and the longest at 2,079 feet, B&G tunnel No. 2 was straight and 754 feet long, and B&G tunnel No. 1 was 682 feet long and curved down to Dry Fork Canyon, totaling 4,795 feet of tunnels on the Bingham & Garfield line.

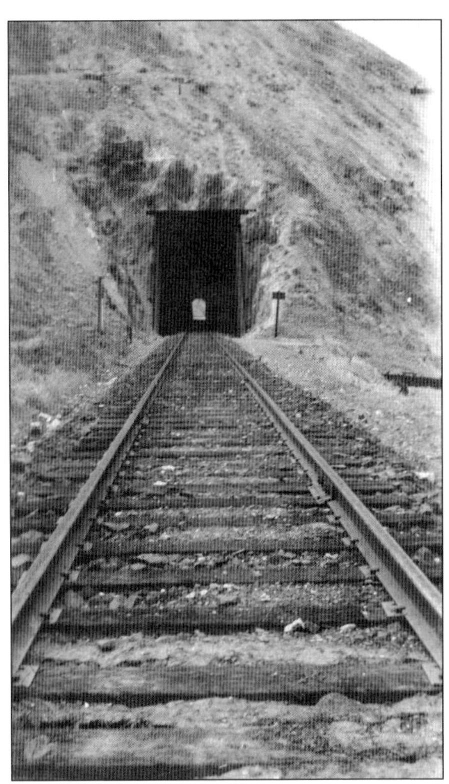

Bingham & Garfield locomotive No. 102 is heading back to the mine tender first, meaning it was traveling backwards (Utah Copper did not have turning facilities at the mine). This enabled the locomotive to travel in the right direction for the downhill trip to the mills. The No. 102 was an Alco-Schenectady 0-8-8-0 mallet-type locomotive, one of four that came to the mine in 1911–1912. The Dry Fork Bridge was the first of the three large steel viaducts on the Bingham & Garfield line; it was curved at 670 feet long and 185 feet high.

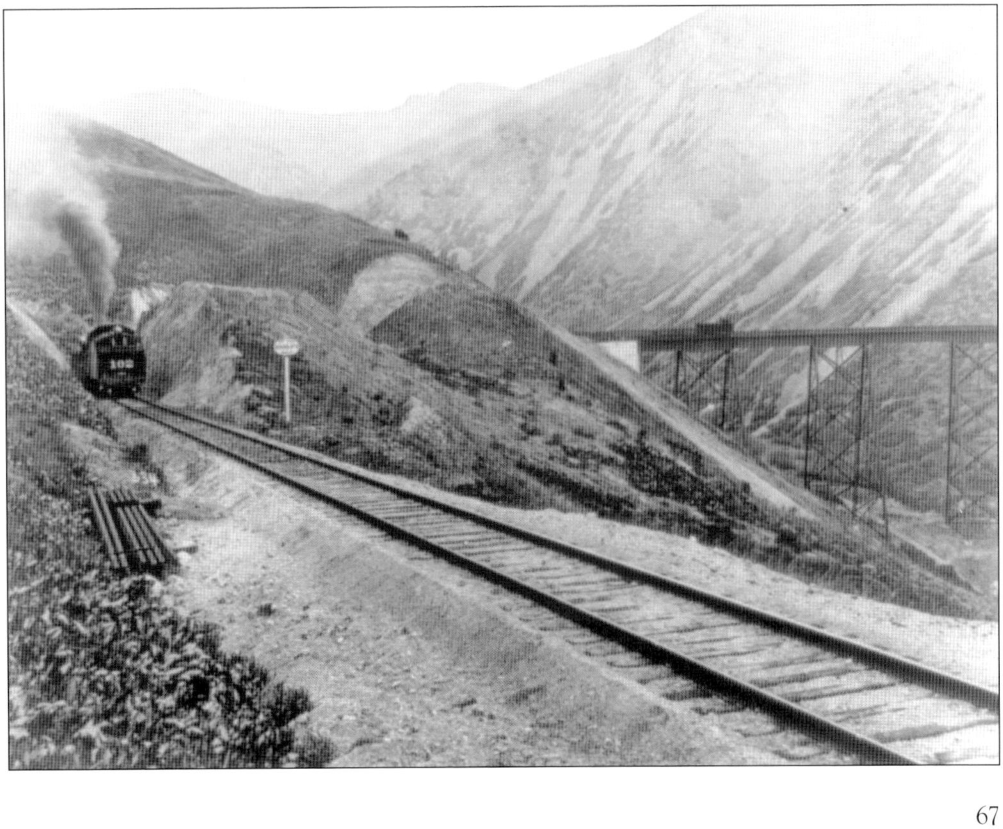

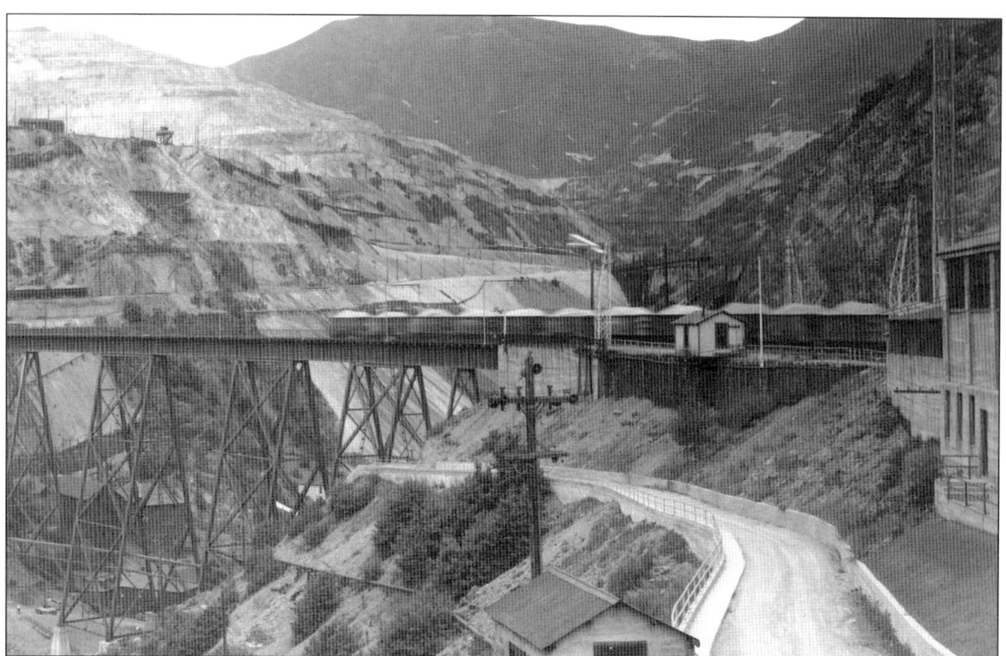

This image shows Carr Fork Canyon as a train rushes over the Carr Fork, or A-line, Bridge. The bridges of Carr Fork started to be built after Utah-Apex mines granted surface rights to Utah Copper to use its property for the dumping of waste rock (overburden). All the bridges except for the A-line bridge were used for waste trains. The Utah Copper open-cut mine is to the left.

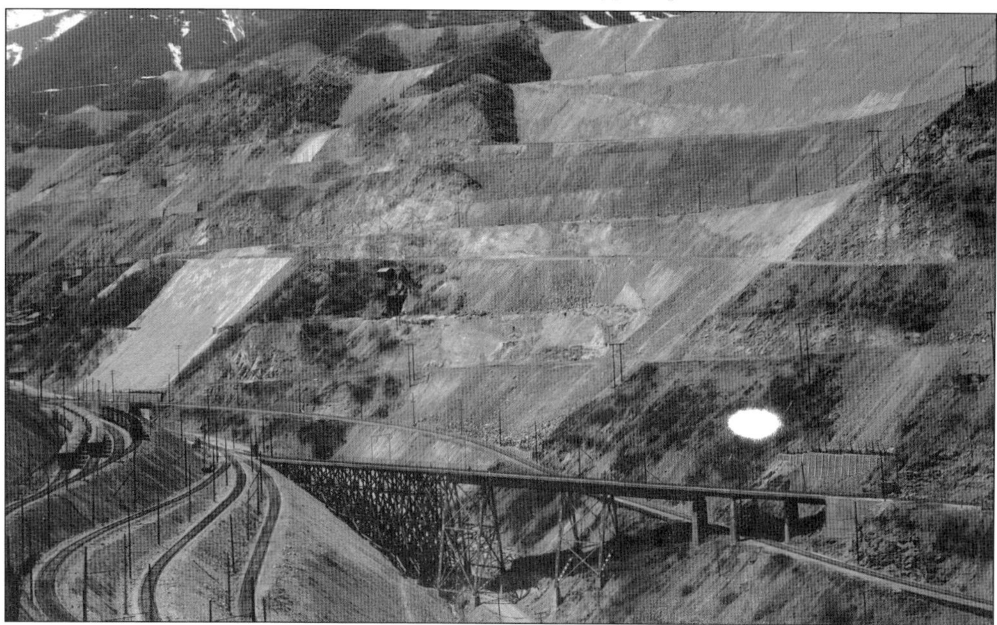

The second bridge crossing over Carr Fork Canyon was the D-line bridge. It was the only straight bridge that crossed Carr Fork, and the construction was different—one half was a wood trestle type, and the other half was built of steel. Older photographs of its construction, dated June 6, 1926, provide the builder date. The large fill area that is seen top right and down to the bridge was once Cottonwood Gulch, now filled with waste rock.

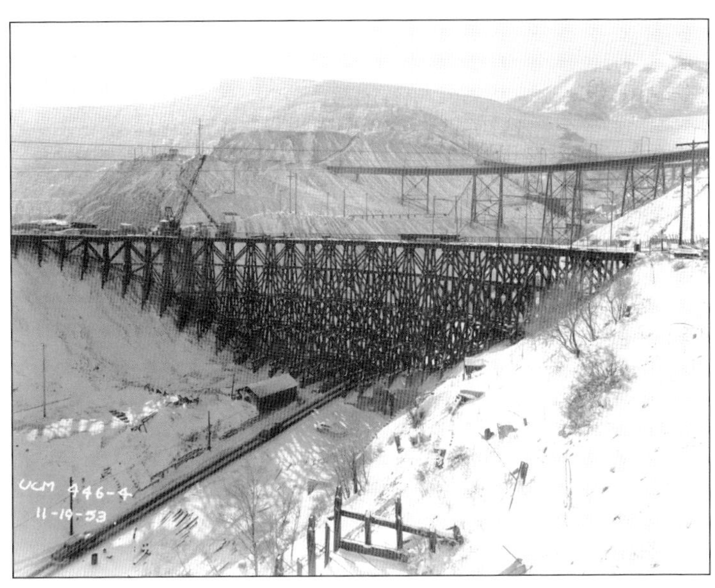

Large cranes work on the south end of the G-line bridge on November 19, 1953. The G-line bridge was built in approximately 1926 as a curved wooden bridge. The bridge was curved because the contour of the mine curved around the mountain that separated the mine from Carr Fork Canyon. Once the mountain was dug away, the south end had to be straightened, and work on the bridge was completed on March 10, 1954.

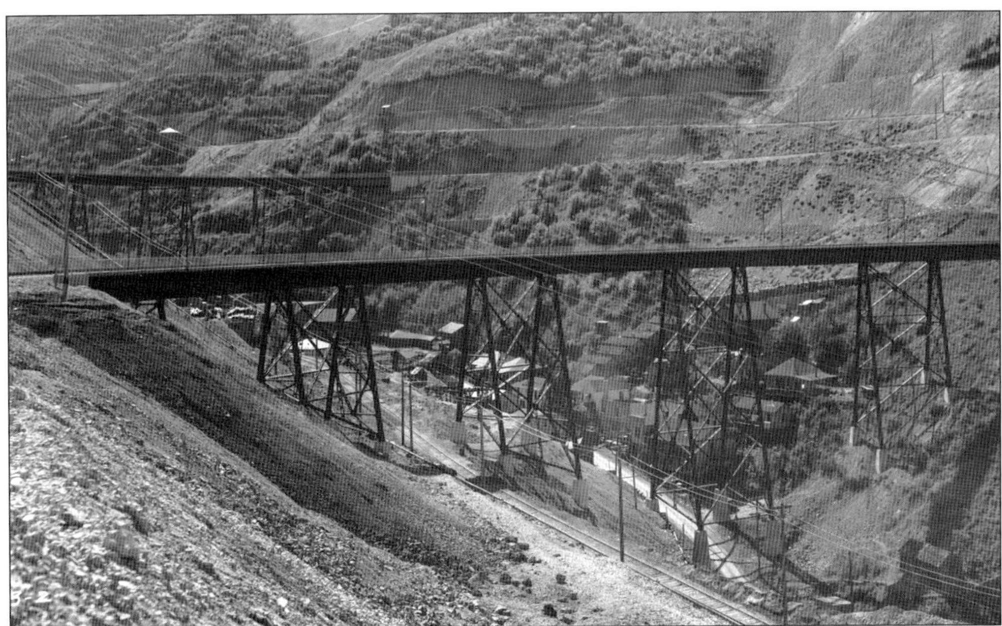

This image shows both H-line and I-line bridges. Both were built in 1939 and are 140 feet high. The H-line bridge was 610 feet long, and the I-line bridge was 590 feet long. All the higher bridges up Carr Fork carried waste trains to the north-side dumps. The stripping ratio at the mine was three to one, meaning three tons of waste rock had to be removed to gain access to one ton of ore.

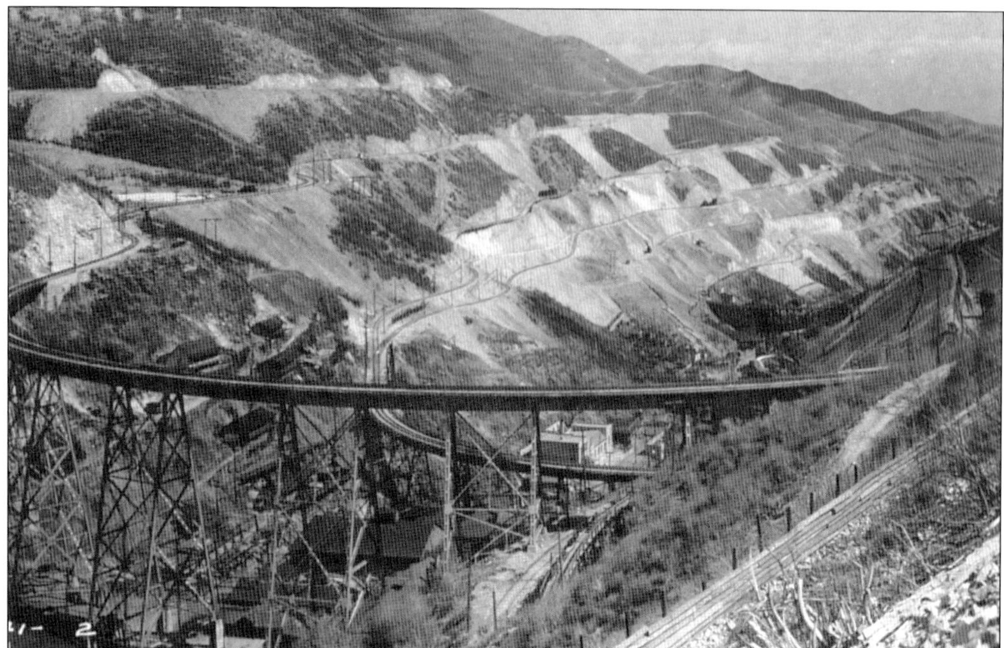

The J-line bridge can be seen through the bridge supports of the larger L-line bridge. The H-line and I-line bridges have not been built yet, so this photograph was taken before 1939. The new Highland Boy elementary school has been rebuilt, as seen at the bottom center. A 1932 fire demolished the school and devastated the town, destroying everything in its path one-third of a mile down the canyon. The fire started just below the J-line bridge.

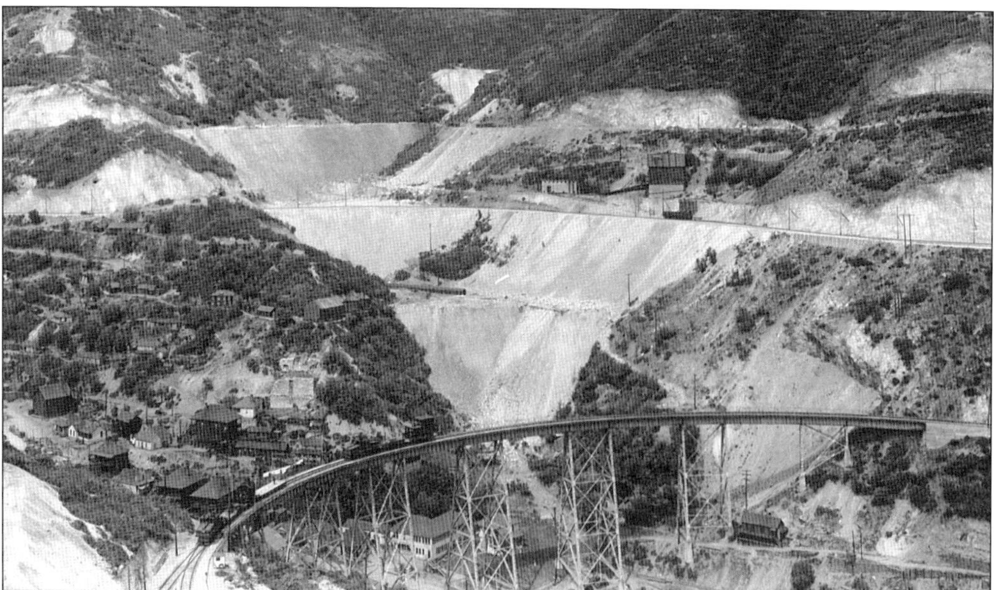

This is another view of the L-line bridge, this time looking up Carr Fork Canyon. The L-line bridge was the last large viaduct up Carr Fork. It was made of curved steel, built in 1926, and was also known as "the silver bridge" because of its color. In the center of the photograph is Sap Gulch, now filled with waste rock. In 1926, an avalanche came down Sap Gulch. A total of 39 people lost their lives, and 15 were injured.

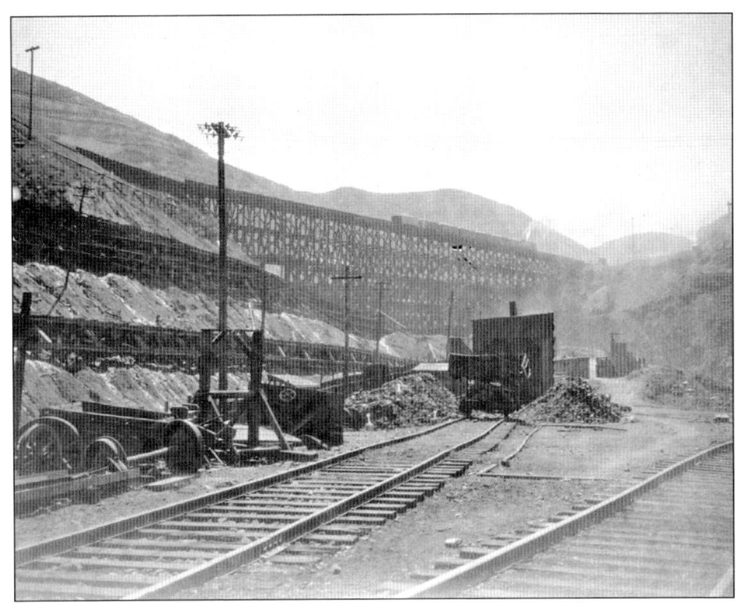

The E-line bridge in Upper Bingham (the name would change to Copperfield in 1914) was on the southeast side of the copper mine, and the bridge separated the town from the mine for a long time. Construction was completed on the E-line bridge in 1914; it was a straight wooden trestle bridge and replaced the old, curved E-line extension bridge, which was built in 1907. The E-line bridge was dismantled in late 1941.

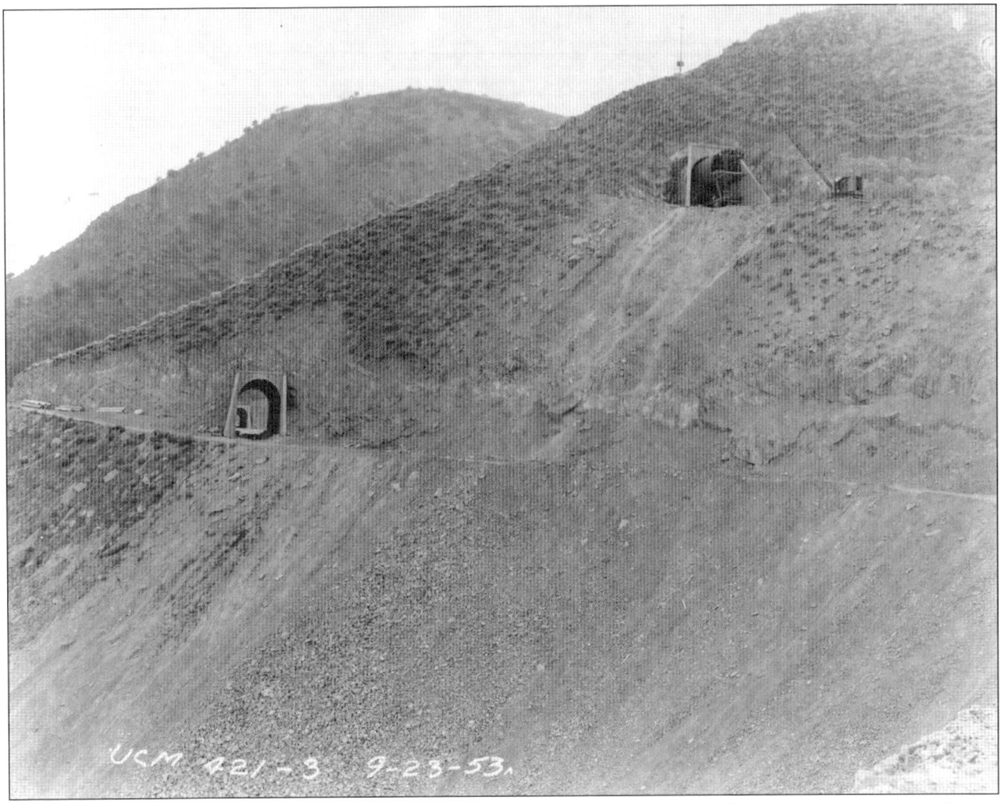

Tunnel construction to access the north-side dumps is seen top right. The waste trains passed over the higher Carr Fork bridges and filled in every side canyon and gulch with overburden. Once the canyons were filled, waste trains had to work their way farther down Bingham Canyon to the next available canyon and gulch. Tunnels had to be built to reach these lower canyons. Utah Copper (Kennecott) was always looking for more room for waste rock.

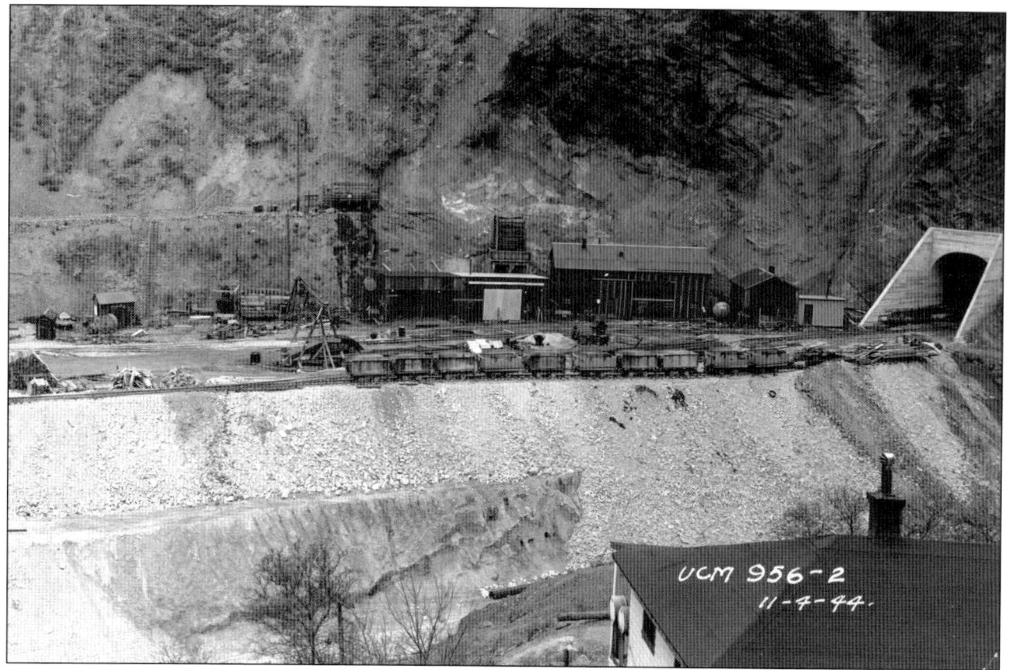

This November 4, 1944, image shows the construction of the 6040 tunnel exit portal. In the early 1930s, once the mountain was starting to be eaten away, the mine was getting deeper, going below an elevation of 6,340. As the mine got deeper, train cars had to be pulled up out of the pit. The electric locomotive could only pull approximately four to six loaded cars at one time.

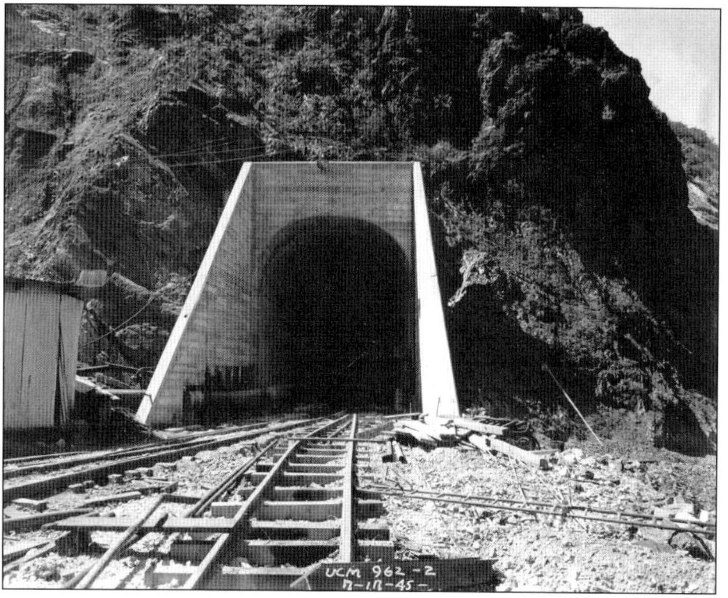

The 6040 train tunnel exit portal is seen here on July 17, 1945. It was the first of three train tunnels to be built in the mine to keep the ore moving downhill. The Utah Construction Company started working on the tunnel on January 10, 1944, and it was completed for operation in February 1946 at a cost of $1,222,000. The tunnel was 4,600 feet long, 18 feet wide, and 22 feet high. The tunnel came out at Frog Town (Lower Bingham).

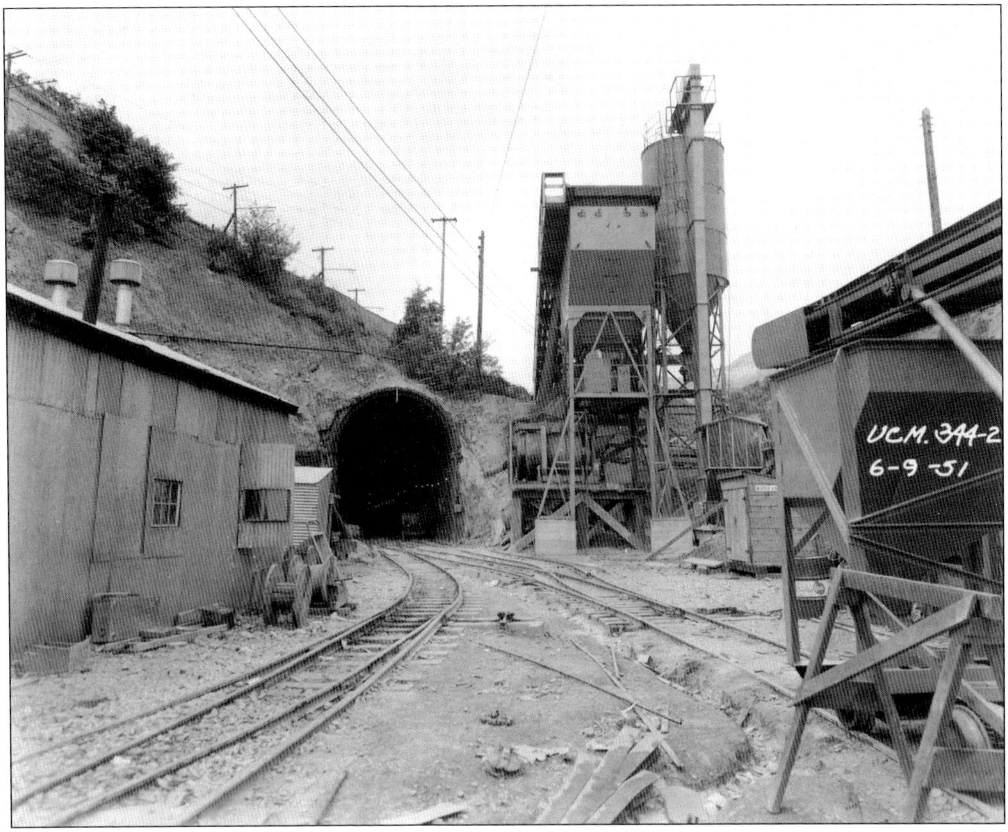

The exit portal of the 5840 train tunnel is pictured here on June 9, 1951; it was the next tunnel in the pit to be built. The 5840 tunnel was 7,042 feet long, 21 feet wide, and 25 feet tall, and construction started in 1951. The exit portal came out in Frog Town by Adderley and Nichols Garage. When building the tunnel from the pit side, the digging was 50 feet below the current bottom of the mine.

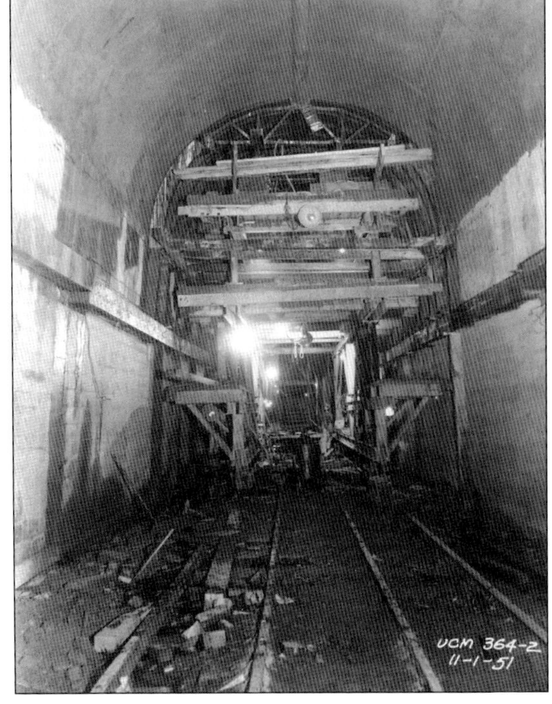

This image, dated November 1, 1951, shows the inside construction of the 5840 train tunnel; note the multiple-level scaffolding. Both the 6040 and the 5840 train tunnel exits came out on the south side of Bingham Canyon in Frog Town. To meet up with the Bingham & Garfield Railroad on the north side, Utah Copper built tunnels in the center of the canyon and filled in around them—this would be known as the cross-canyon connection.

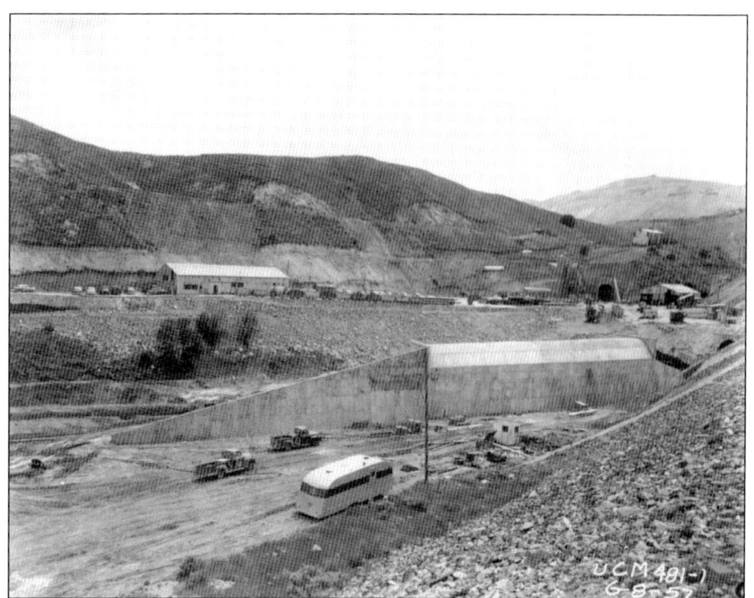

This image shows the last train tunnel exit portal that came out of the mine, center right. The bottom tunnel elevation was 5,490, thus its name. The cement tunnel under construction would be filled in around it so the trains coming out of the 5490 tunnel could pass over it and enter Copperton yard. The cement tunnel was the route of the Denver & Rio Grande Western to Frog Town.

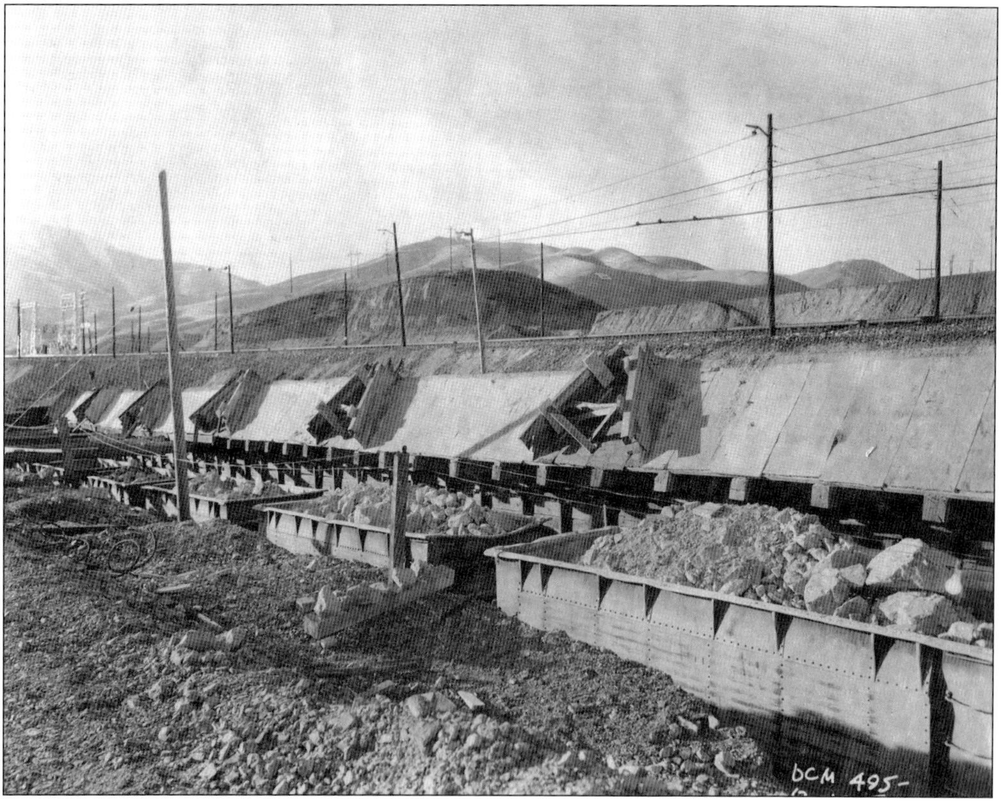

This loading station was built when ore was found while digging the 5490 train tunnel. The 5490 train tunnel path went through the US Mining Company property, so when construction crews found ore, it was dumped into standard-size ore cars and delivered to its smelter in Midvale. The tunnel was 18,000 feet long, the longest train tunnel in Utah. The 5490 train tunnel housed the conveyor belt and went into operation in January 1988.

Seven

Haul Trucks Take Over the Mine

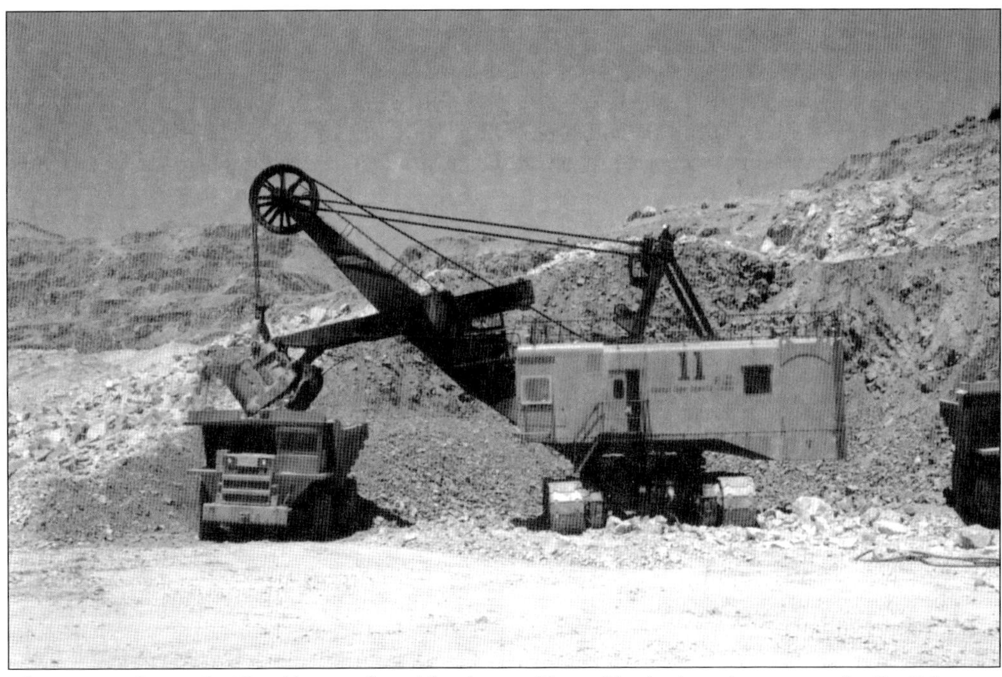

This image shows the No. 11 rope shovel loading a 65-ton Haulpak with waste rock. On February 23, 1963, Kennecott announced a $100 million mine expansion, which included the top two-thirds of the mine using large-haul trucks instead of its railroad. Trucks were more versatile and could make a wider level pass quicker and more cost-effectively. Most of the top portion of the mine was waste rock removal.

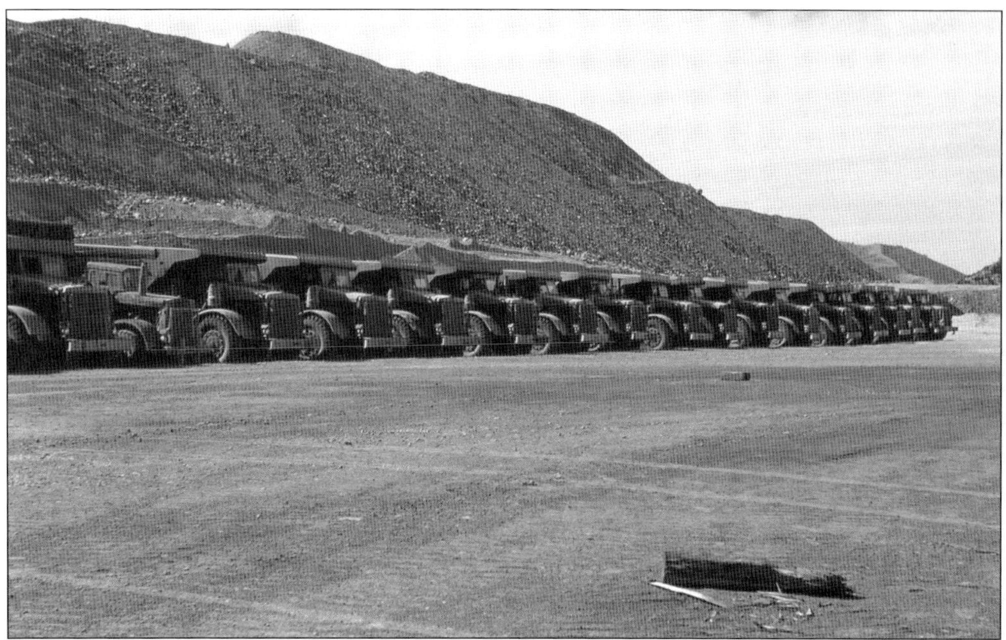

Pictured is Western Contracting's fleet of 50-ton Euclid dump trucks at Kennecott in August 1960. Kennecott contracted with Western Contracting back in 1958 to remove waste rock in the higher levels at the mine. Over the next five years, Western Contracting removed additional millions of cubic yards of overburden. Kennecott stated that it was not practical to use its railroad system on the upper levels of the mine.

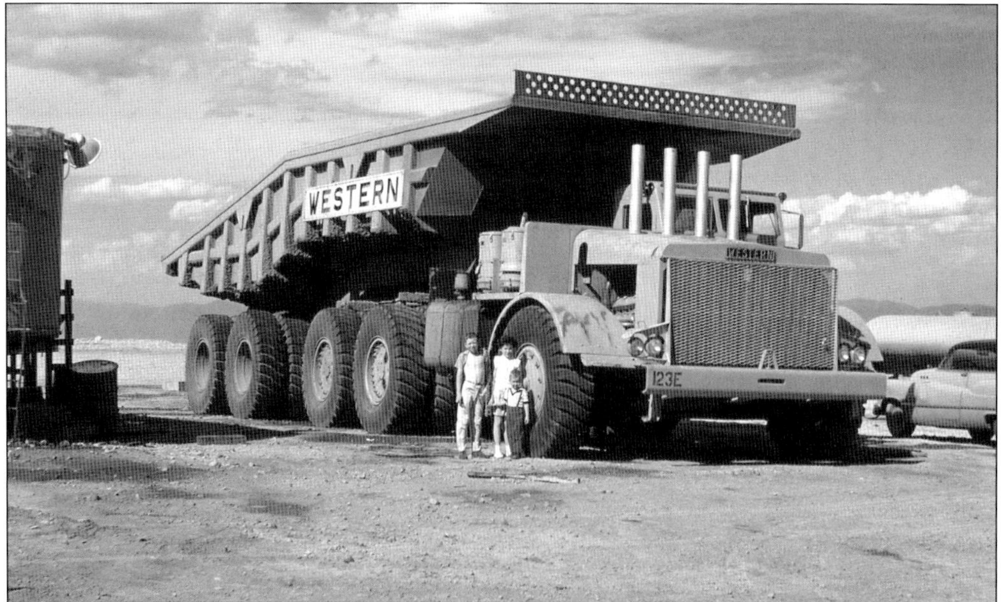

Western Contracting brought what was called "the World's Largest Truck" to Kennecott in 1958. It could haul 100 cubic yards of earth, carry a net payload of 165 tons, and had a 750-horsepower engine. The huge truck stands 14 feet high and is 15.5 feet wide and 55 feet long. When the dump bed is lifted, it raises 45 feet in the air. The children are those of Shirl Scroggin, head of security at Bingham.

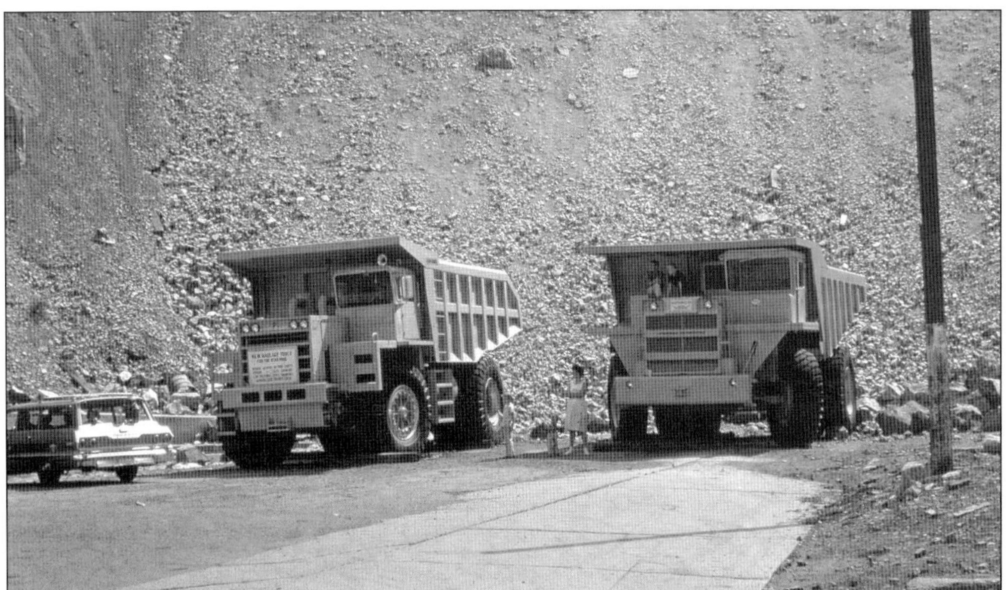

This image shows two new haul trucks on display at Bingham Canyon in September 1963. Both are 65-ton trucks; the one to the left is a KW Dart (the "KW" was a shortened name for Kenworth), and the truck to the right is a Haulpak by WABCO (Westinghouse Air Brake Company). The first haul trucks were small at the mine but would continually get larger. The Shirl Scroggin family poses for the photograph.

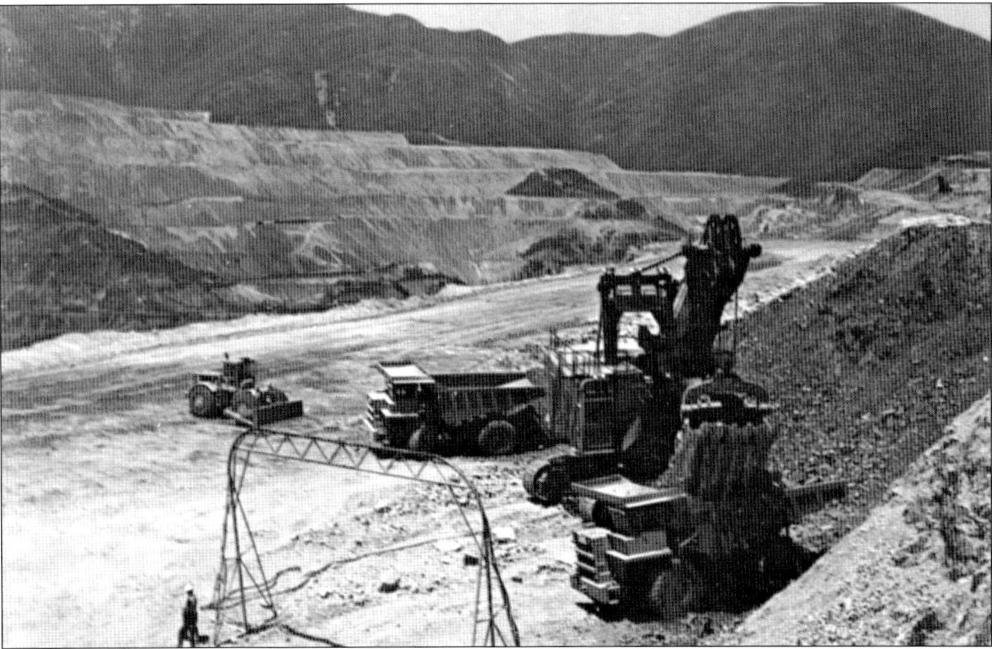

This image shows two 65-ton Haulpaks; one is being loaded, and the other truck is in position to be loaded. The rubber-tired dozer is used to clean up overspill. This image shows how large the truck levels are compared to the train levels. Work is being done on the higher portion of the mine to remove waste rock. The cable bridge is how the electric shovel received its power. Electricity ran through what was called a trail cable; the bridge prevented damage from the large-haul trucks.

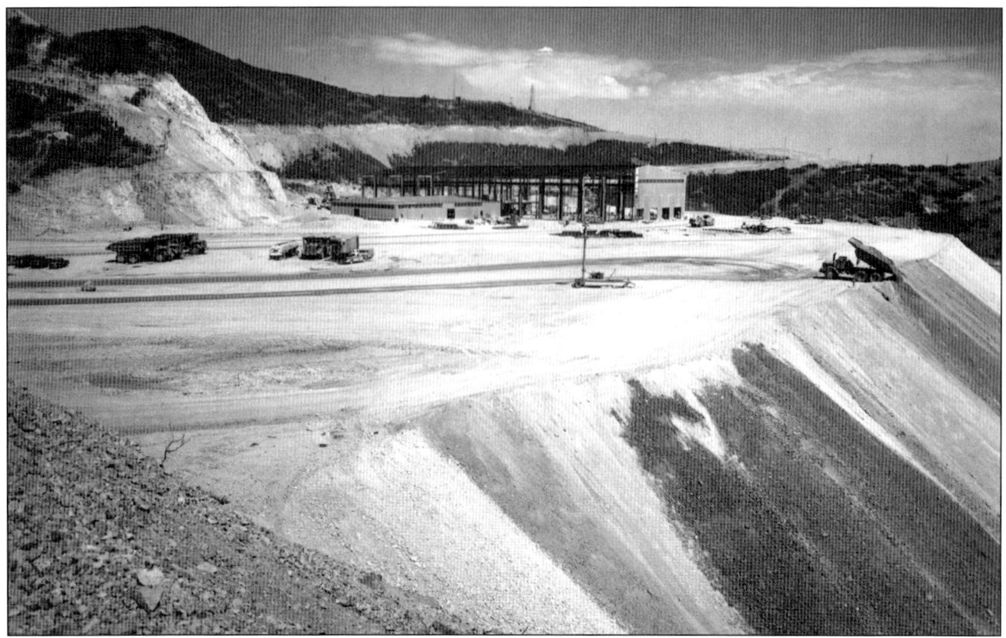

Part of the $100 million mine expansion was to build a large truck shop for repairs and maintenance for Kennecott's new truck haulage. The Yosemite truck shop was named after an underground mine that was once at this location, on the east side of the Oquirrh Mountains. Construction started in 1963, and it opened in 1964. Note Western Contracting trucks at work dumping on what will be Yosemite Dump.

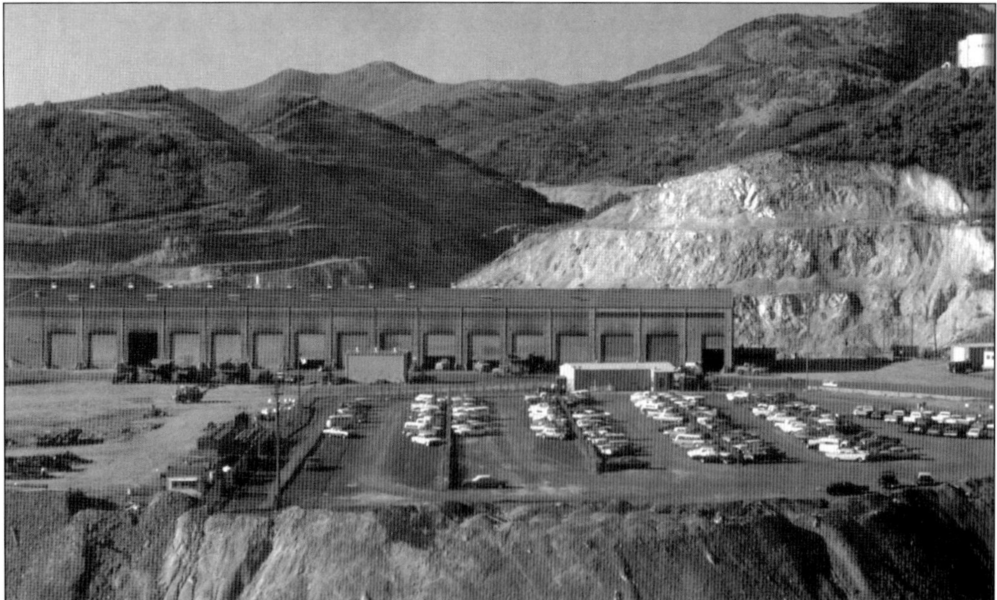

This image shows the Yosemite truck shops in December 1970. Soon after this photograph was taken, every other door in the shop had to be replaced with a larger door to accommodate the new, bigger haul trucks. By March 1965, there were a total of 79 haulage trucks in use at Bingham, including 65-ton Darts and Haulpaks, 85-ton Lectra Hauls, and 110-ton Dart tractor-trailer trucks. The Yosemite shops were buried in 2003.

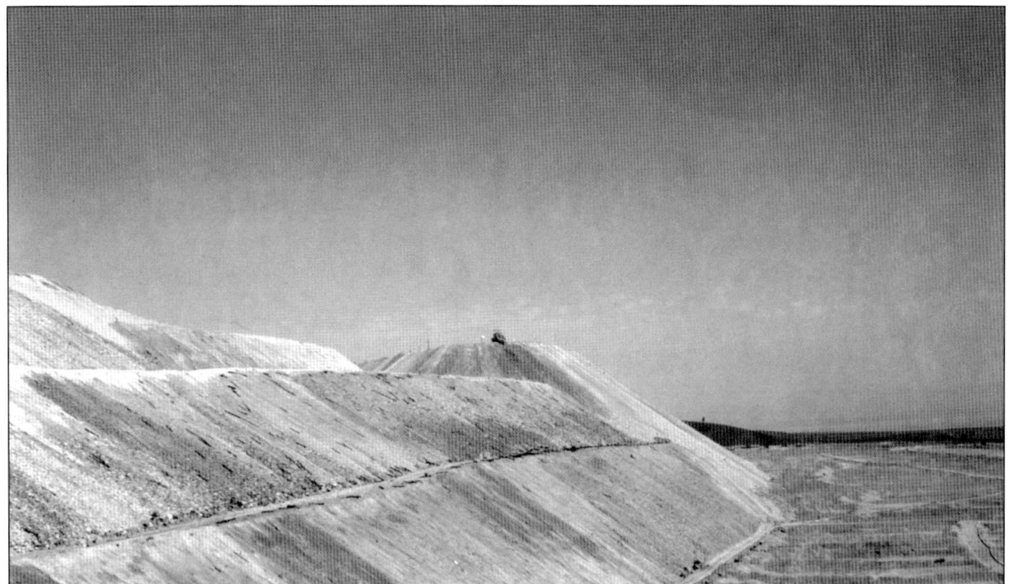

This 1970 image shows a haul truck dumping waste rock on a new dump called "code 51." All the dump sites around the mine had code numbers; this was to keep track of where each truck was working and how many loads they made a shift. The new code 51 access road went down Bingham Canyon, high above the canyon floor. The new dump was started above the old railroad dumps, as seen in the bottom right.

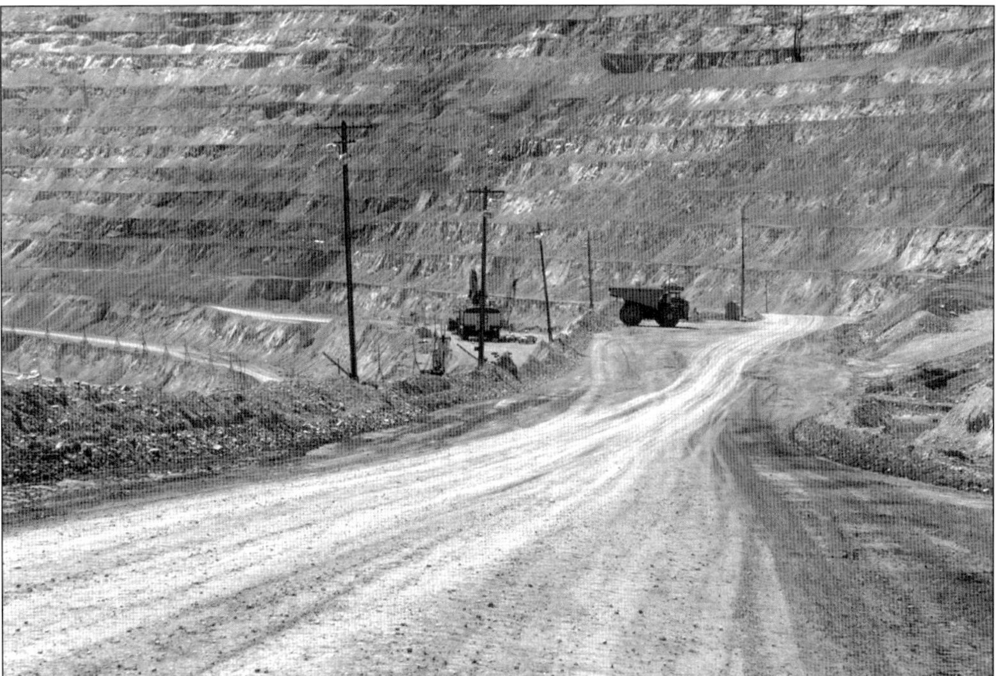

This image shows a haul truck sitting at a reloading station. The highest elevation that the trains worked was 6,240, and the haul trucks operated in the top two-thirds of the mine. When ore was found in the higher portion of the mine, haul trucks would work their way down to a reload station and dump their loads, then a shovel on the level below would reload the load into a railroad ore car.

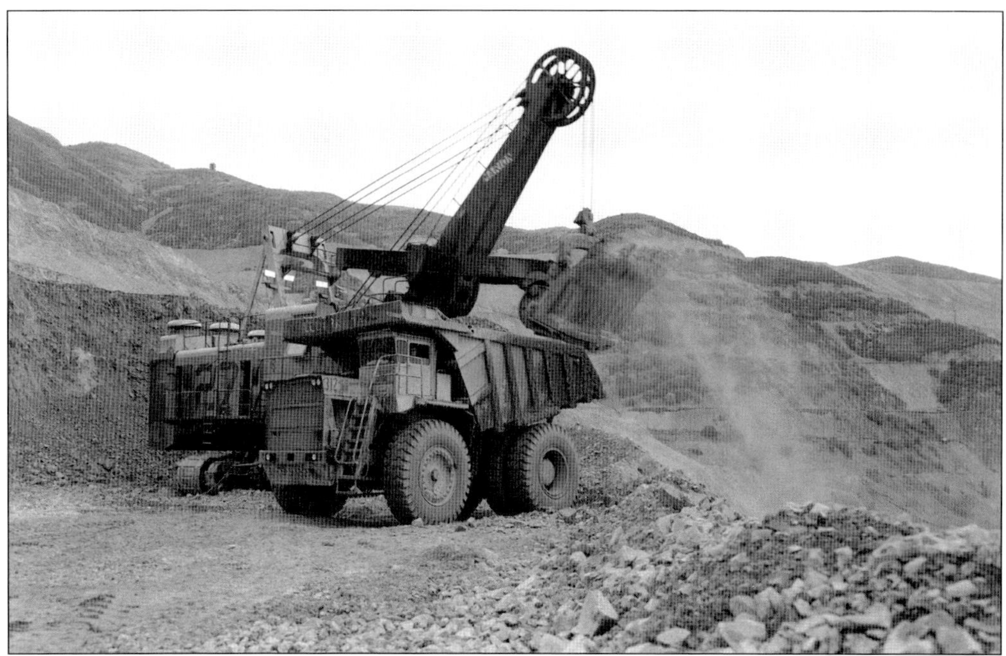

The No. 312 haul truck is being loaded by the No. 27 electric shovel. The haul truck is a 100-ton Lectra Haul by Unit Rig. It has electric drive motors in the back wheels, while the truck's engine was the generating plant for the wheels. Not long after truck haulage started at Kennecott, the trucks got bigger. The first trucks were the 65-ton trucks, but larger truck development was going on.

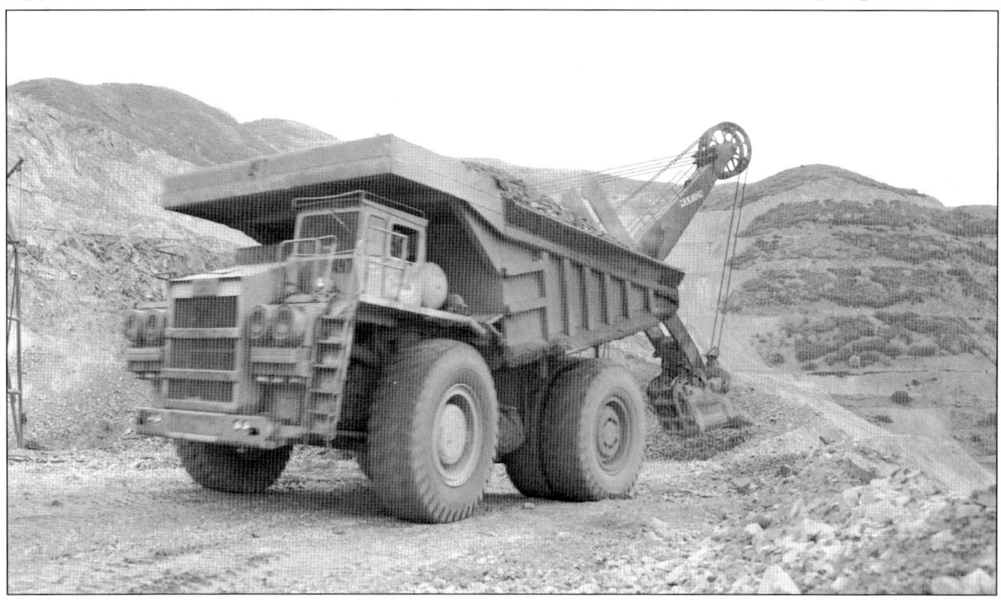

Pictured here is the same place and shovel as the previous image, but this haul truck is No. 497, a 100-ton WABCO Haulpak. WABCO was also working on General Electric's motorized wheel drive. In the 1960s and 1970s, companies like WABCO, Unit Rig, Terex, Dart, and Euclid were competing for the large-haul truck business. Kennecott was a testing ground for some of these trucks, both diesel-electric and mechanical drive trucks. The company was even testing turbine engines in the trucks.

This image shows the new No. 11 electric rope shovel at the mine on September 10, 1963. The No. 11 shovel is a Marion 191-M, 15-yard shovel. With the haul trucks getting bigger, the electric shovels also had to get bigger. With the arrival of two new shovels in the summer of 1971, Kennecott had a total of 37 shovels: 9 fifteen-yard, 1 twelve-yard, 6 eight-yard, 5 seven-yard, and 16 six-yard shovels.

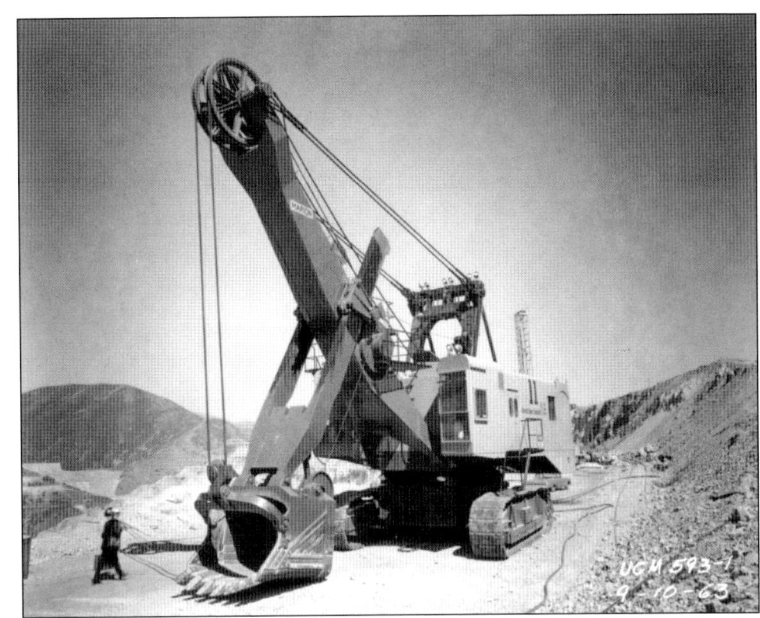

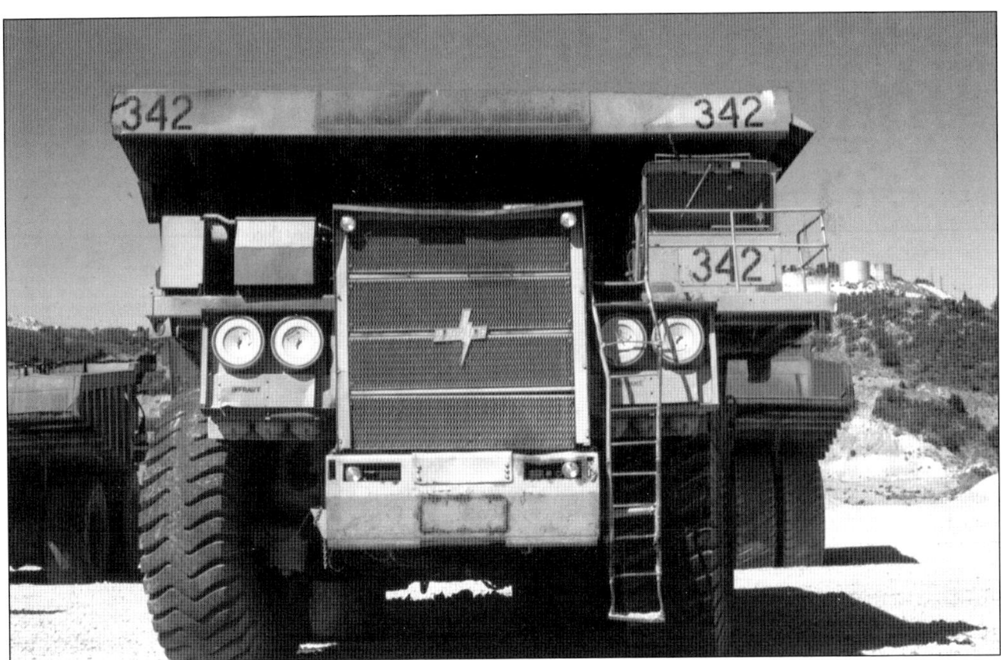

By the time this photograph was taken in the early 1980s, the 342 Unit Rig Lectra Haul had seen a lot of miles. In May 1973, twenty-one new 150-ton Lectra Hauls came to the mine. The trucks were assembled at the old Lark mine shops. The trucks had a 1,600-horsepower engine to power the generator for the rear electric drive motors and a 500-gallon fuel tank; the tires were 10 feet tall.

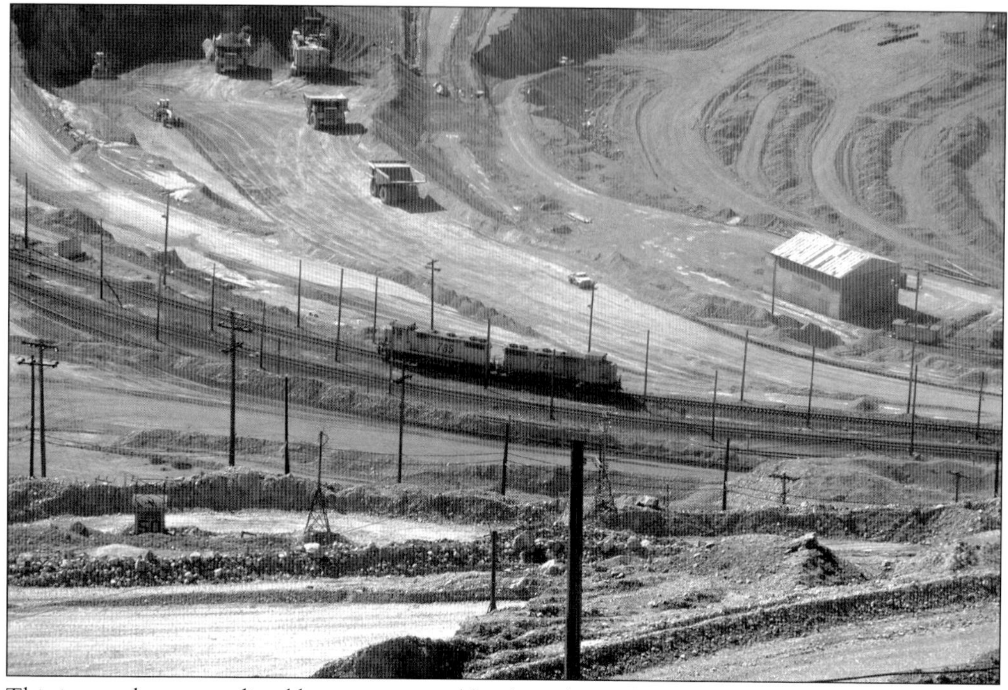

This image shows two diesel locomotives and haul trucks working in the lower mine. In 1983, the entire mine went from a shovel operation to a truck mining operation. Trains would only be used at reload stations located at the three tunnel levels: 6040, 5840, and 5490. In 1983, Kennecott's fleet of trucks was twenty 150-ton Euclid, eighteen 150-ton WABCO (Haulpak), and fifty-nine Unit Rig (Lectra Haul).

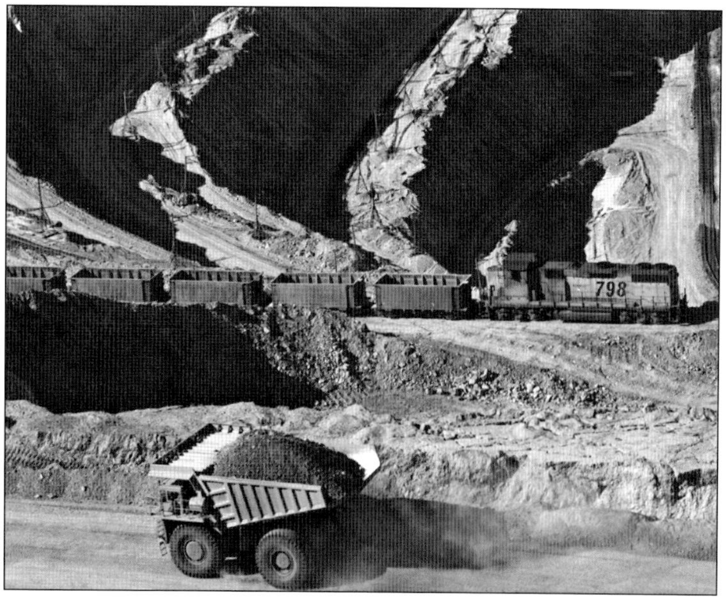

This view shows a haul truck and train working together at the mine. In the early 1980s, the ore was still being hauled by trains to the mills from reloading stations. Note the railroad track and the many catenary towers still around the levels for the once electric railroad. The towers also brought electricity to the shovels; their three wires were on the opposite side of the catenary wire. All would be removed soon.

This September 4, 1984, photograph shows haul trucks traveling on the new road that went down the north side of Bingham Canyon. Once it was decided to make the whole mine a truck operation, a centralized location was needed, and it became the old 6190 area. A new road was built down the north side of Bingham Canyon that would eventually lead to the main access road into the mine, named the "ten percent."

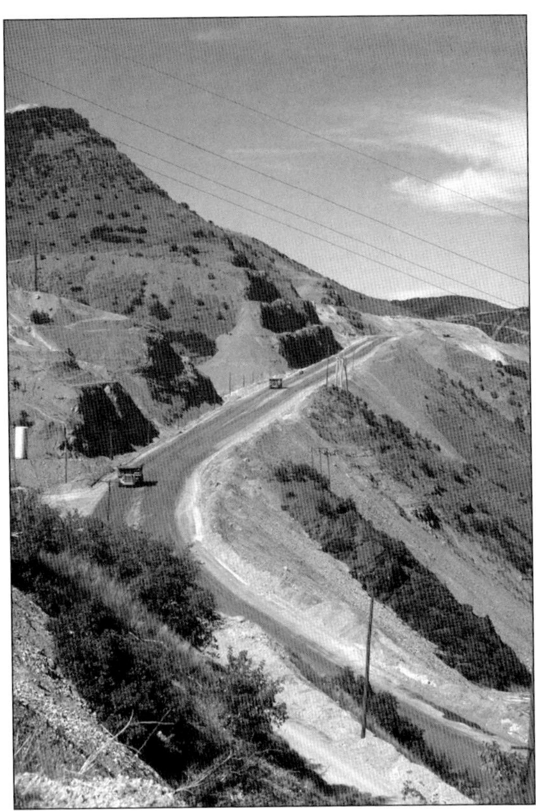

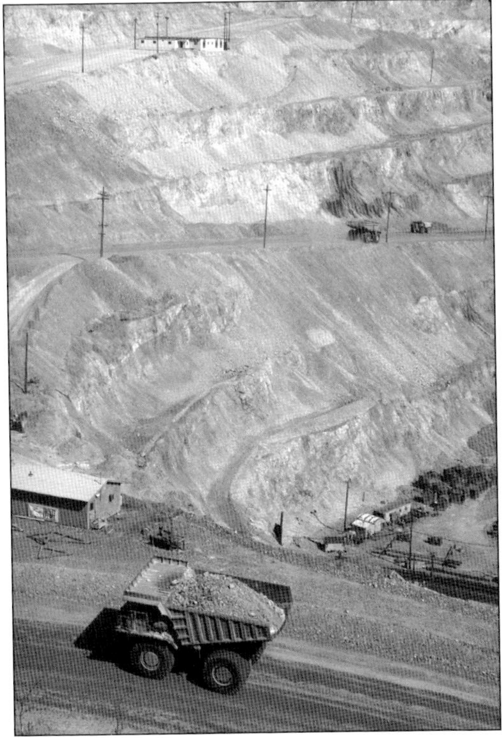

This image shows a haul truck moving on the north side of the mine, bottom left. An empty haul truck and a water truck are about to pass on the main access road (the ten percent), center right. The mine would expand as far as it could to the north. Note the tracks to a reload station, the track office building, and the many small electrical switch shacks stacked together, bottom right.

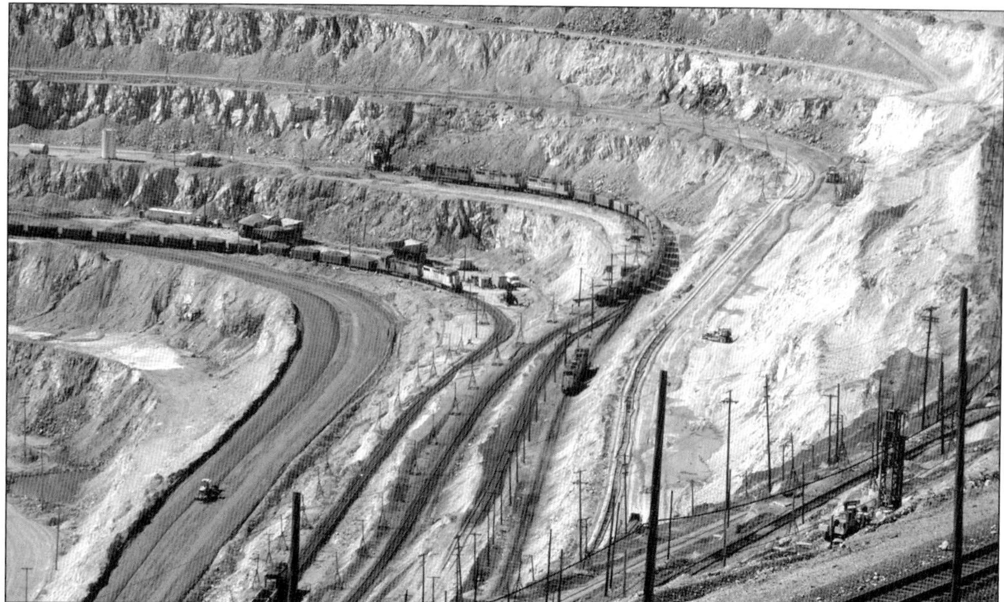

This image shows trucks and trains working together in the pit on August 24, 1983. The 1980s were turbulent years for Kennecott, through buyouts and takeover; the mine changing hands several times; Standard Oil, Kennecott's parent company, taking charge; and British Petroleum's large interest. In 1989, Rio Tinto purchased all the assets for $4.3 billion. The mine was named Rio Tinto Kennecott.

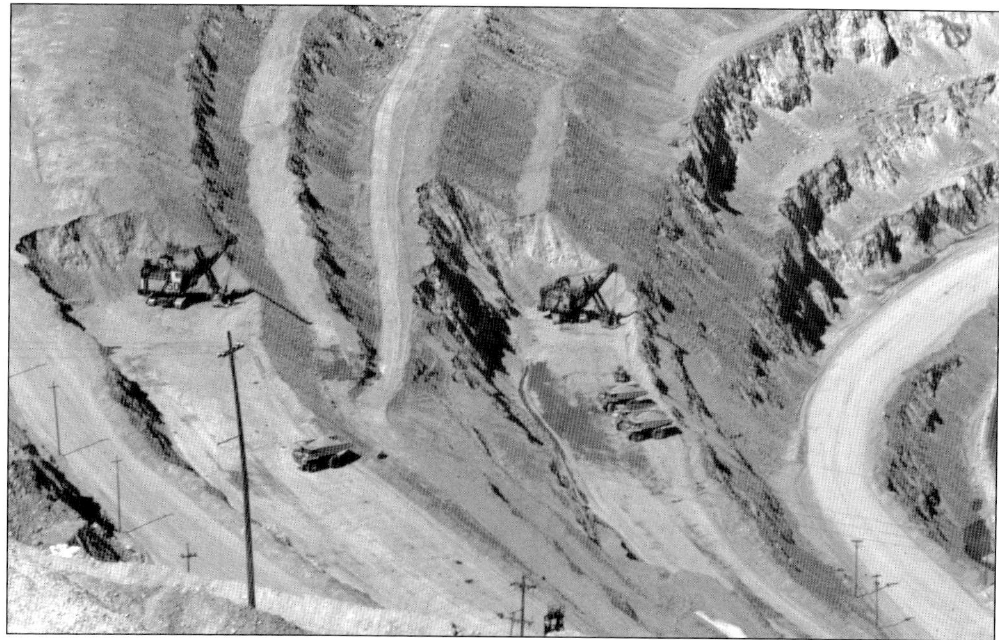

This September 1984 photograph shows trucks taking over all mining operations. On March 26, 1985, Kennecott announced that it would spend $400 million to modernize the Bingham Canyon Mine. That included a new Copperton Mill, a new in-pit crusher with a conveyor system to deliver ore to the mill, and an upgrade to all haul trucks, shovels, and mining equipment. When the modernization was completed, the workforce was 1,800, compared to 7,300 hourly employees in 1980.

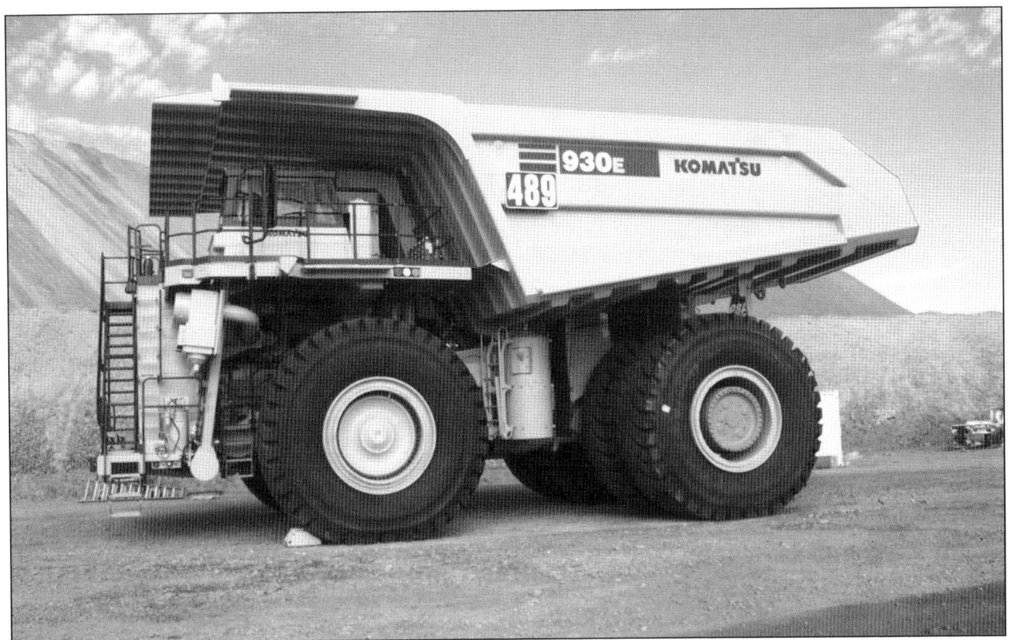

A new Komatsu 930E haul truck comes to the mine. The Komatsu 930E truck had a payload of 320 tons and was 26 feet and 7 inches wide, 24 feet high, and 50 feet long. The truck was the first to use AC motors for its electric wheel drive. Trucks at the Bingham Canyon Mine in 2019 totaled 107 and included 84 Komatsu 930E, 5 Caterpillar 794AC, and 18 Caterpillar 793D.

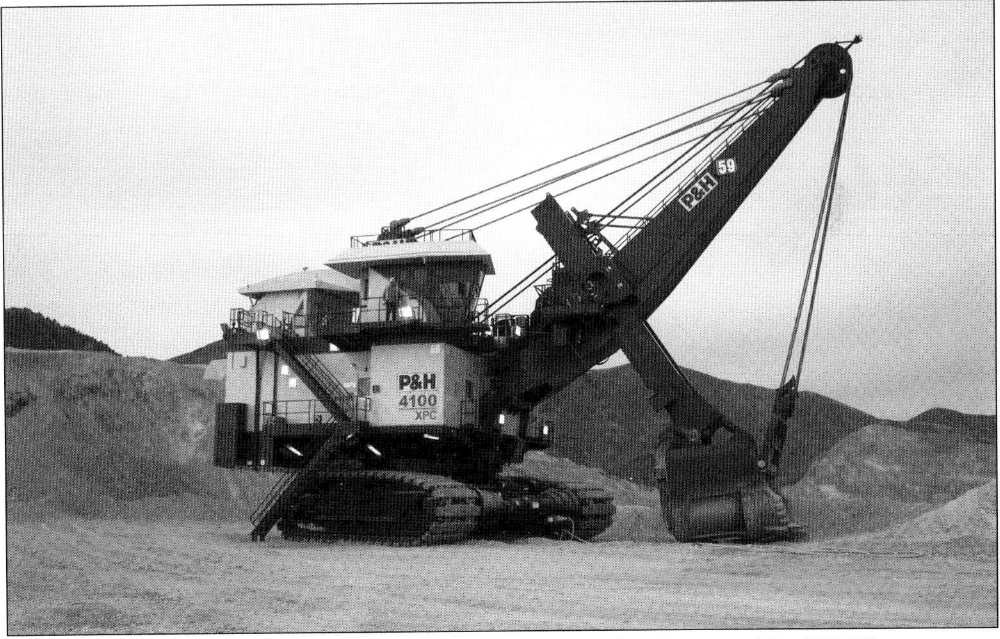

This image shows a new P&H 4100XPC electric rope shovel at the mine. The "P&H" name came from its founders, Alonzo Pawling and Henry Harnischfeger, in 1884. The company grew over the years, developing mining equipment. During the truck-developing years at Kennecott, the shovel equipment grew as the trucks got larger. Shovels at Bingham in 2019 included three P&H 4100XPC 72-yard, three P&H 4100 54-yard, and two P&H 2800XPC 45-yard shovels.

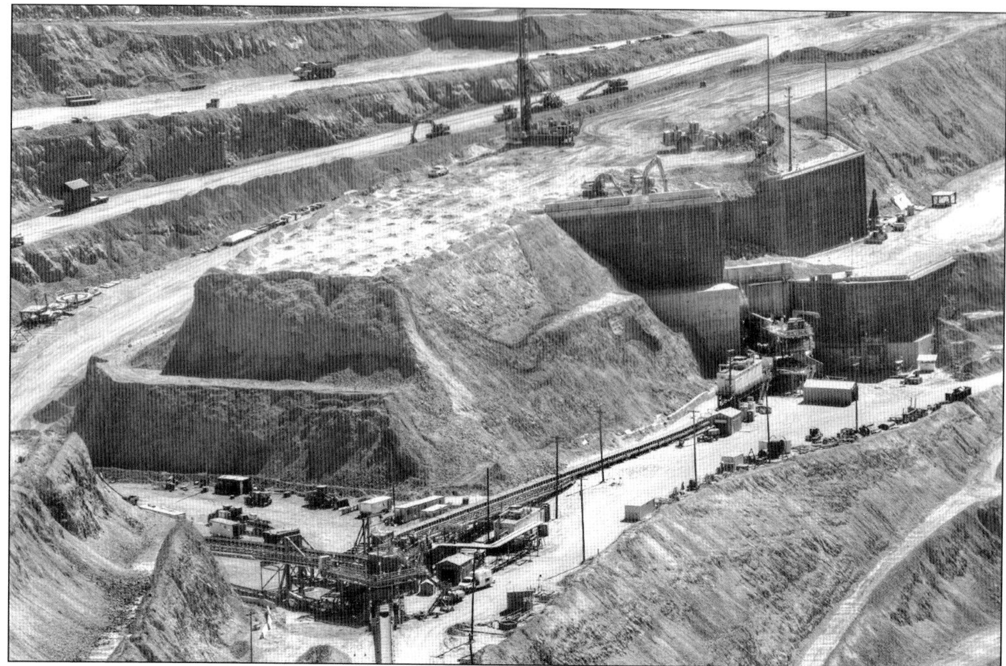

This image shows the in-pit crusher site being moved on September 27, 2011. The in-pit crusher and conveyor system were first put into operation in January 1988. The conveyor was installed inside the old 5490 train tunnel. The longest of the six separate conveyor belts was 17,300 feet in length. When the conveyor belt was replaced due to normal use in August 2002, the conveyor had carried 700 million tons of material.

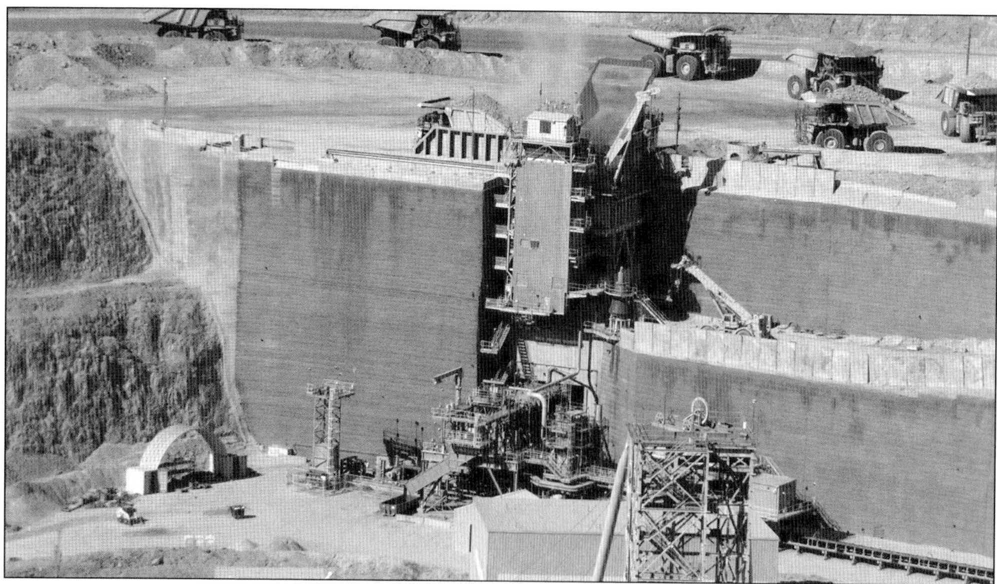

Large-haul trucks line up to dump at the crusher on October 26, 2024. The crusher and conveyor system moved out of the mine in 2021, and the new location is at the old 6190 area. The crusher's original location was by the 5490 train tunnel to use the tunnel for the conveyor system. The first move of the crusher was in 2002 to accommodate mine expansion to the east, and the second move was in 2011.

Eight

Diesel Locomotives

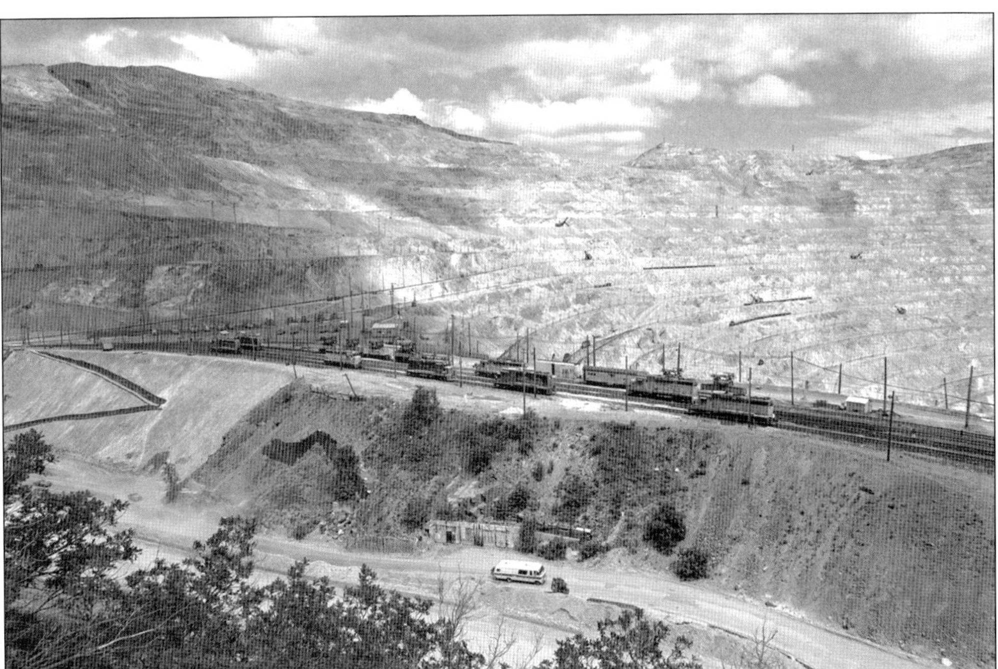

This 1976 James Belmont photograph shows diesel locomotives at the 6190 yard. The electric locomotives gave Kennecott years of great service, but many were getting old; several of the older electric locomotives came to the mine in 1928 and 1929. In May 1973, the first diesels came to the pit to be tested because the company wanted to see how they performed on the steep grades at the mine. (James Belmont.)

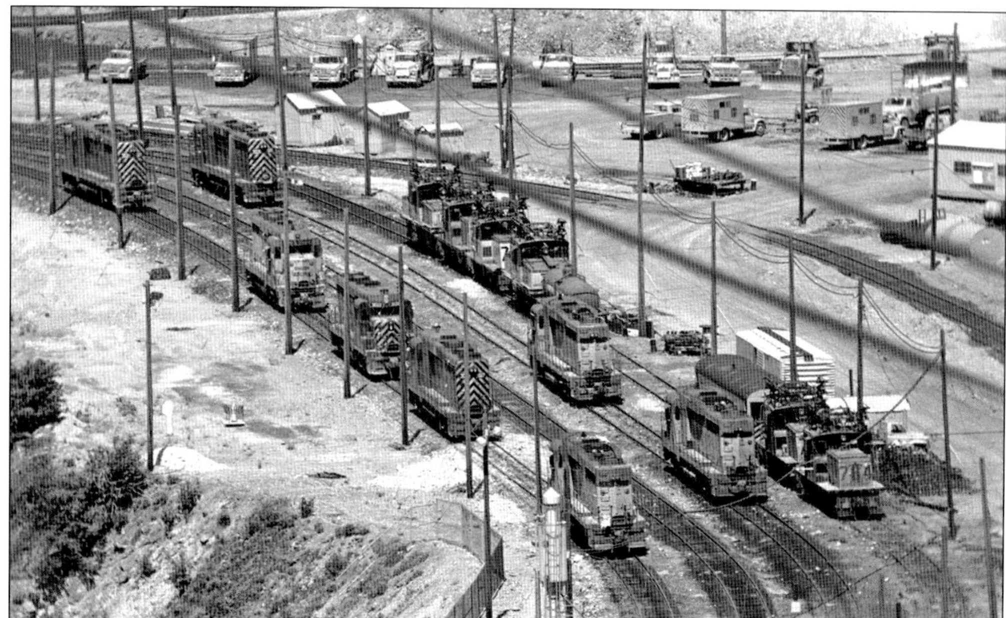

This view shows both steeple cab electrics and diesel locomotives at the 6190 yard. Kennecott leased 12 EMD-30 diesel-electric locomotives from Union Pacific and the Denver & Rio Grande Western Railroad from May 1973 until December 1978 to evaluate their motive power in the mine; the diesel locomotives were used to remove waste rock. Another two locomotives that showed up at the mine were from the Atchison, Topeka & Santa Fe Railway. (James Belmont.)

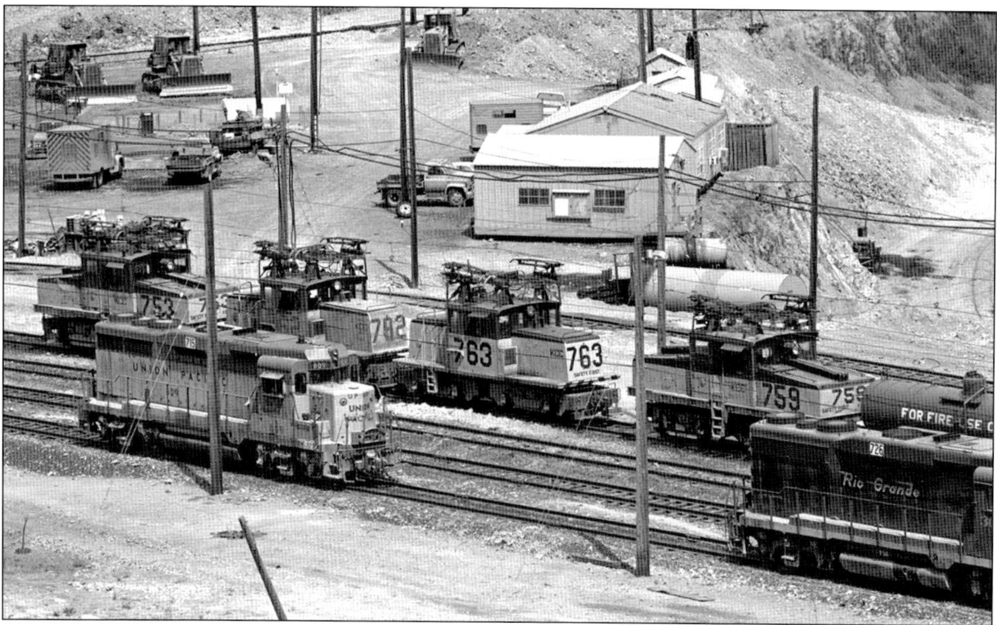

This close-up shows some of the test diesel-electric locomotives that Kennecott used at the mine. The Union Pacific and Denver & Rio Grande Western locomotive has been given temporary numbers. The test results were successful, so Kennecott decided which locomotive would do best at the mine. Kennecott had previously used diesel locomotives at the Magna and Arther mills, the Garfield smelter, and the precipitation plant. (James Belmont.)

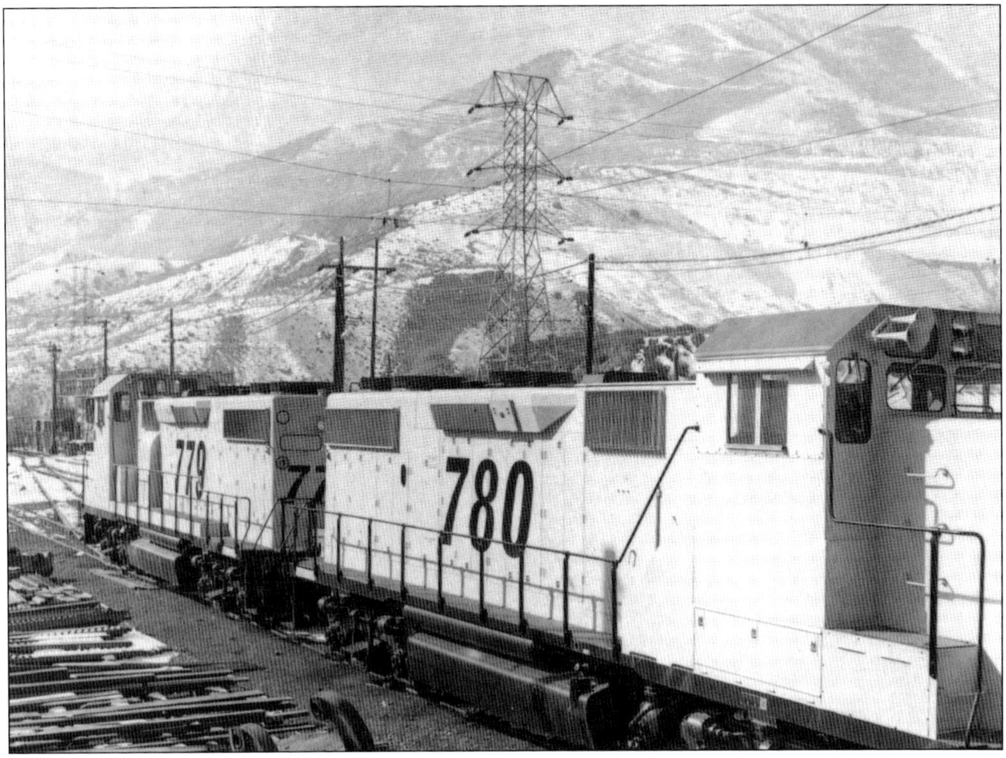

In this February 1977 image, Kennecott's new diesel-electric locomotives, the 779 and 780, are at the Dry Fork train shops. After the extensive testing of the diesel-electric locomotives proved to be successful, Kennecott decided to purchase EMD GP 39-2s from the Electro-Motive Division of General Motors. The new locomotives had specially designed cabs that were 26 inches higher than the main body to visually assist the loading of ore and waste cars.

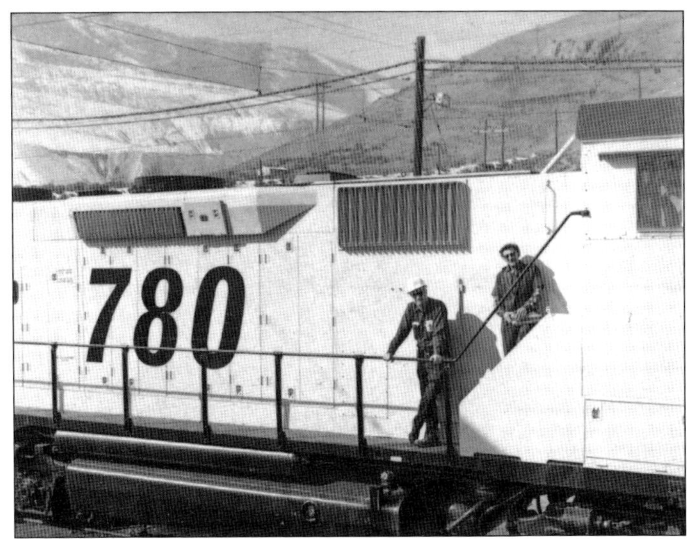

This image shows train general foreman Stan Long and Joe Rakich inspecting the new locomotives at the Dry Fork shops. The first 11 locomotives were delivered in February 1977; they were numbered from 779 to 789, and their builder dates were all the same: January 1977. The new diesel-electrics were 12-cylinder, two-cycle, 2,300-horsepower turbocharged locomotives, 59 feet long and 18 feet high. The new locomotives were purchased to improve efficiency and cost per ton.

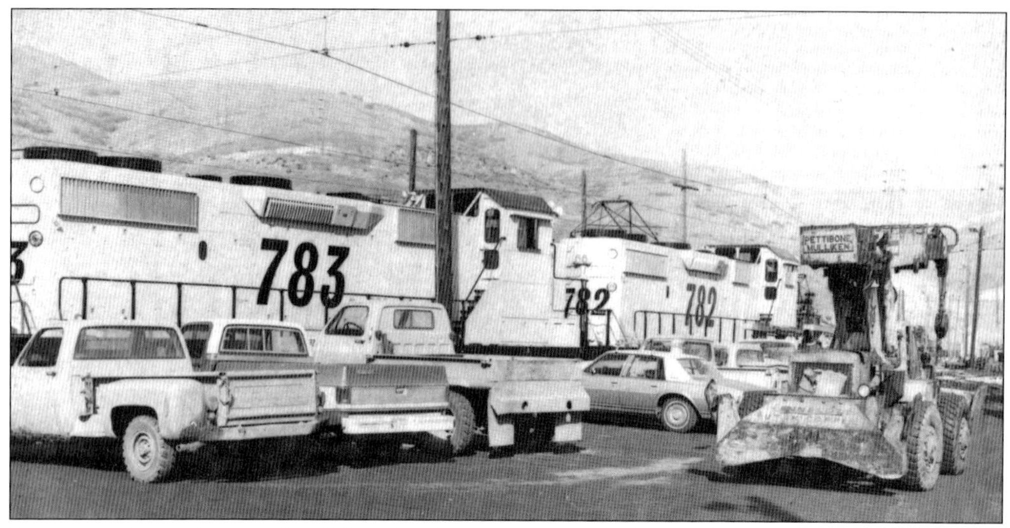

This image shows the new Nos. 782 and 783 locomotives at the Dry Fork shops. The new locomotives had a smaller fuel tank of 2,600 gallons instead of the standard 4,000-gallon tank. The fuel tanks were mounted off the rail higher for rock clearance. The smaller fuel tank allowed room for an extra air reservoir for better braking, and the locomotives were equipped with a dynamic brake for the four-percent downhill grades.

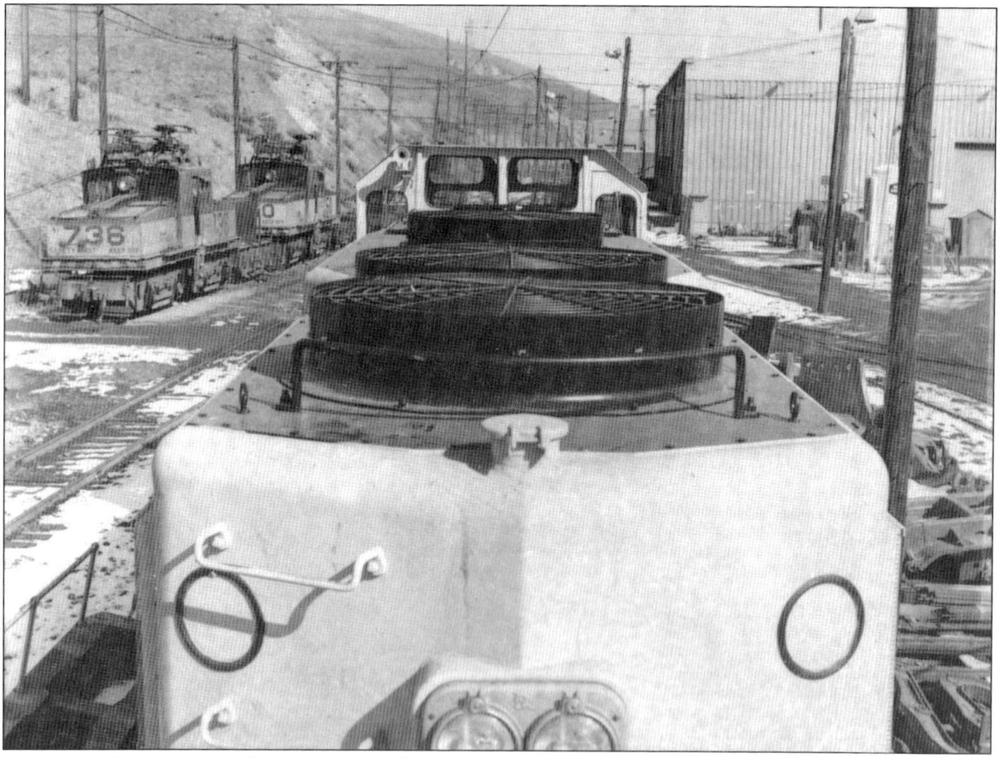

This view shows one of the new diesel-electric locomotives with its specially designed cab, called a "vista cab." The cabs were raised 26 inches over the main body of the locomotive. The cabs assisted engineers while the shovels loaded the ore or waste cars. The shovel operator would hold his dipper out over the train car and drop his load when he wanted the train to stop.

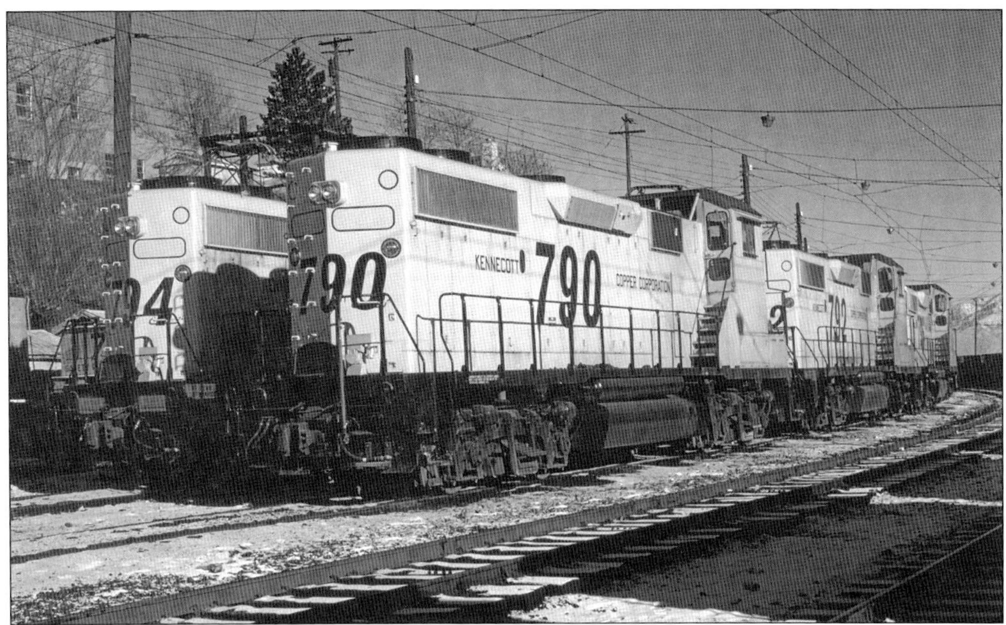

In January 1979, railroad photographer James Belmont captured new diesel-electric locomotives at Copperton yard. In November 1978, Kennecott purchased 10 more locomotives, numbered 790 to 799. All the locomotives were painted yellow with black lettering. The image shows the smaller fuel tanks on the locomotives that are 13 inches off the rails, which provides rock clearance from shovel overspill or spillage by the moving loaded ore cars. (James Belmont.)

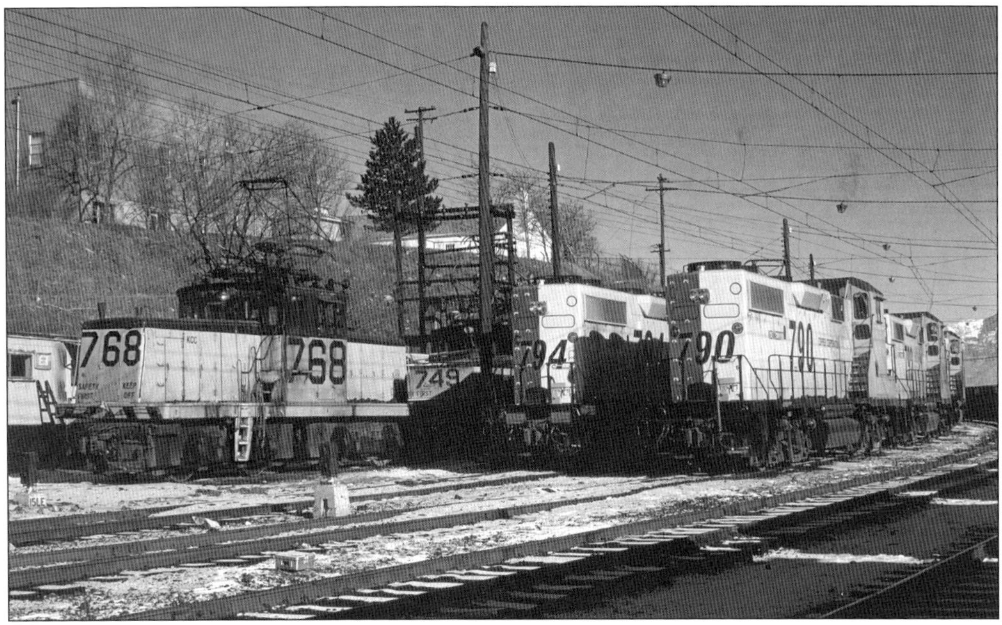

This photograph, taken by James Belmont, shows both the old steeple cab electric and the new diesel-electric locomotives at Copperton yard in January 1979. The No. 768 electric locomotive GE 125-ton was delivered to Lead Mine on March 7, 1952. Lead Mine was the interchange point between the Denver & Rio Grande Western and Kennecott. Locomotive No. 768 was retired from service after 31 years at Kennecott on November 11, 1983. (James Belmont.)

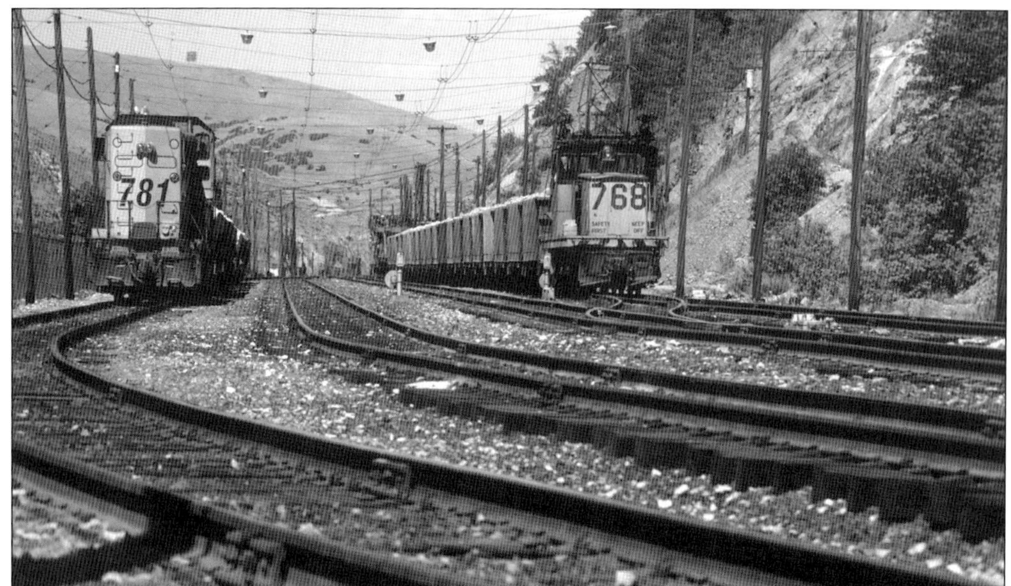

This image shows the No. 781 diesel-electric locomotive hauling a waste train and the No. 768 electric locomotive hauling an ore train through the 5840 yard. The No. 781 locomotive came to the mine on February 11, 1977, with the first group of 11 locomotives. Kennecott first used the new locomotives to haul waste. Waste train trips were usually shorter, consisting of traveling from the shovel to the dump. Note the No. 768 electric motor; it is receiving power through the top pantograph.

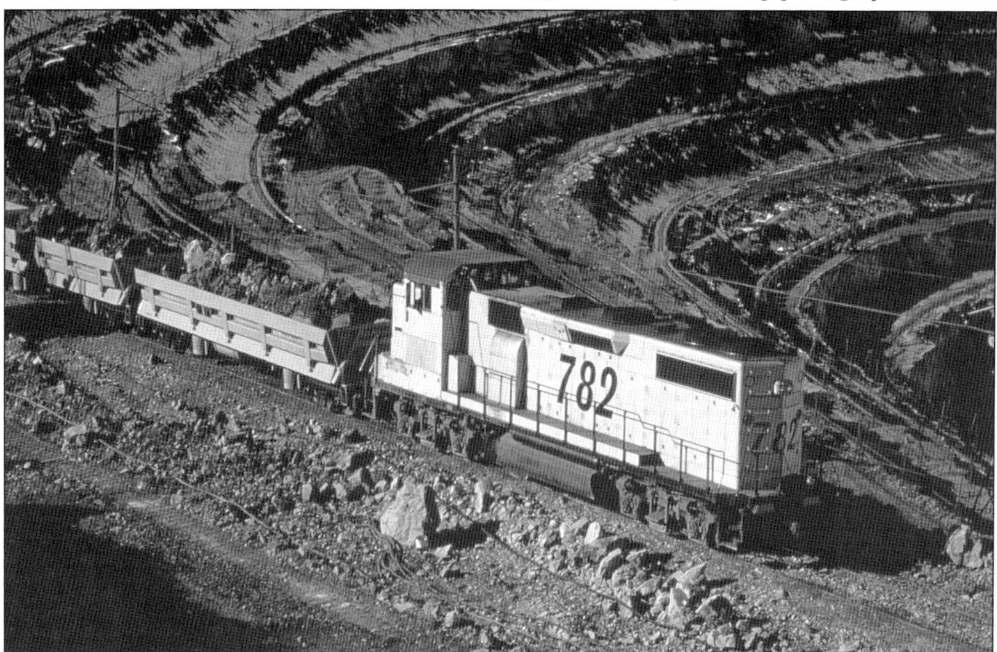

This image shows the No. 782 diesel-electric locomotive pulling waste cars out of the mine. The No. 782 locomotive was one of the first 11 that came on February 11, 1977. Notice the portable towers and catenary wire for the electric locomotives; once they are removed, waste haulage can start using larger electric shovels. The photograph shows the importance of the raised fuel tank, with the many rocks around the mine.

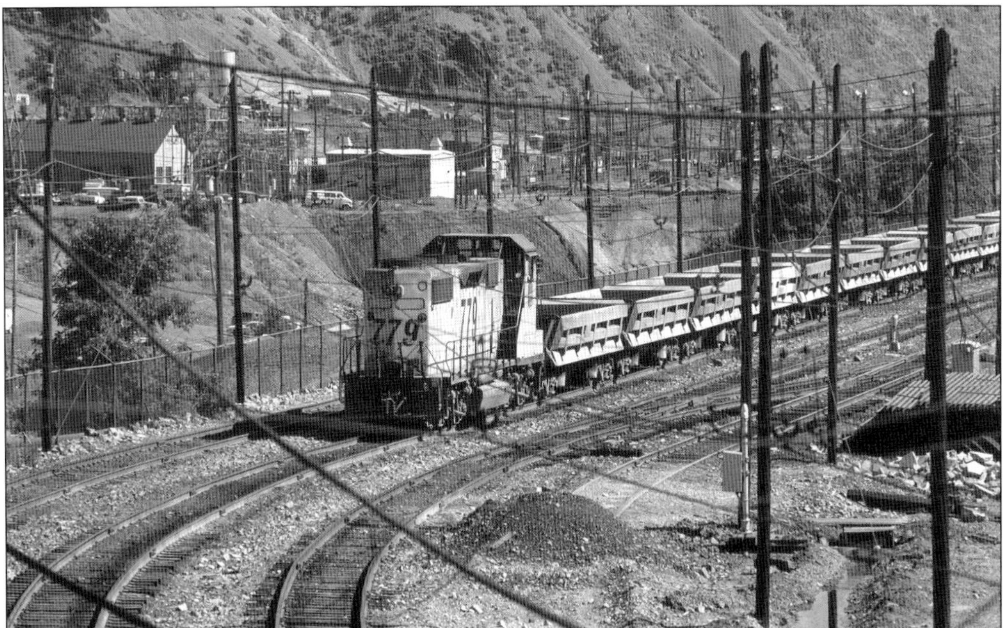

The No. 779 diesel-electric locomotive pulls a long string of waste cars back to the mine. The train travels through the central yard. Diesel-electrics were able to handle more waste cars than the electrics. The last seven of the EMD GP39-2s locomotives came to the mine in late 1980 and were numbered 705 to 711. In 1981, the old steeple cab electric locomotives were being scrapped or donated to museums.

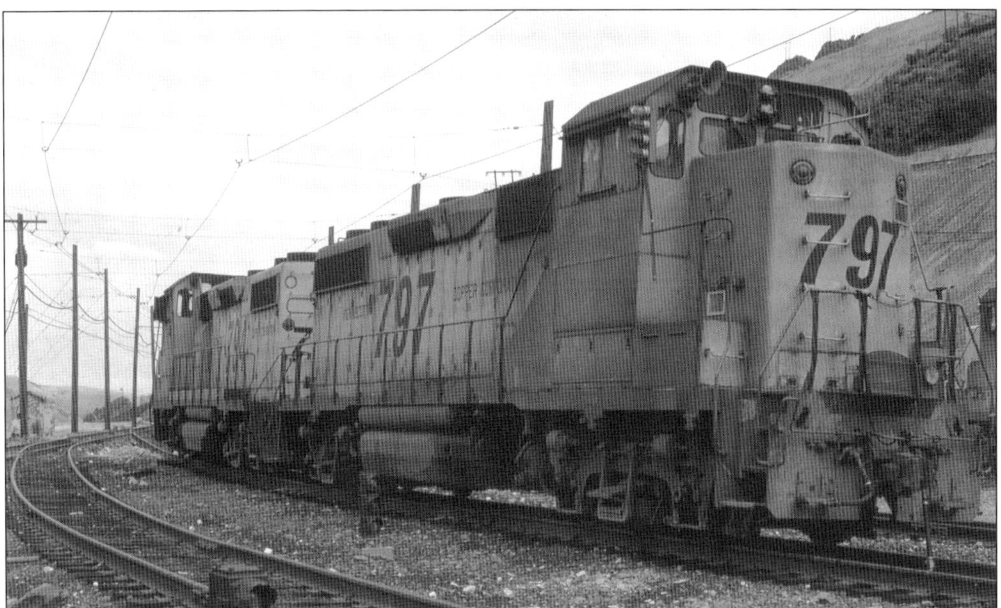

This image shows the Nos. 797 and 794 locomotives at work. Production foreman Gary C. Curtus made this statement: "The new diesels locomotives were a work horse, I think we may have been able to ship eighteen to twenty loads through the 6040 tunnel straight down 5860/6040 switch back and on down to the CC load line with these GP 39-2 locomotives, the dynamic brake really made the difference."

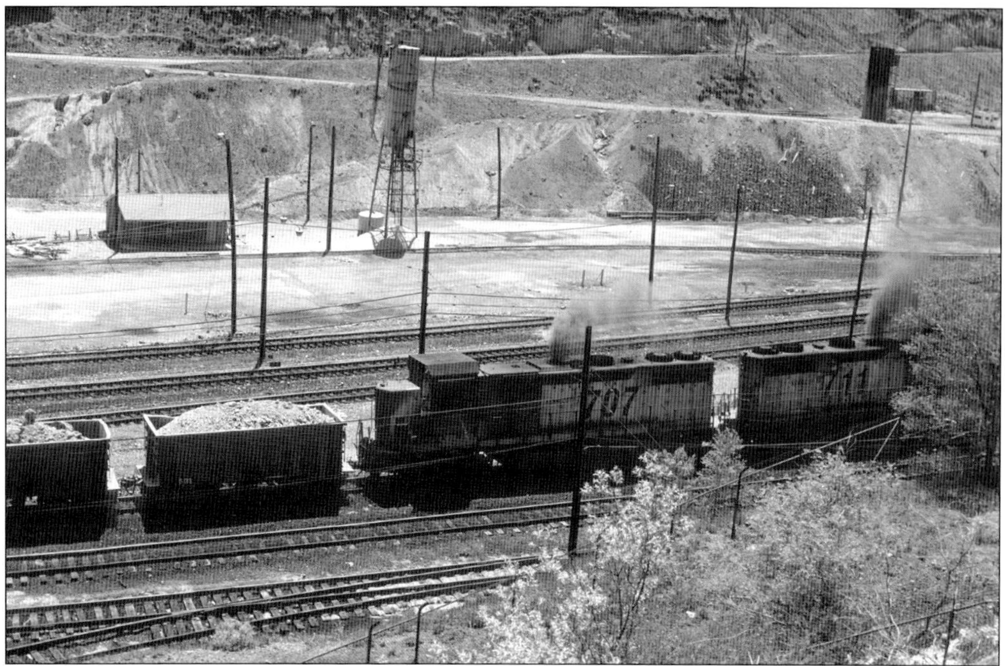

This image shows an ore train coming into Copperton yard. In 1983, trains were only used at reload stations at the three tunnel levels of the mine: 6040, 5840, and 5490. The 5490 reload was removed when the conveyor took over the tunnel. Foreman Ron Burke said, "We really made money when that reload was in. . . . We were up to 50 cars per train there at the end."

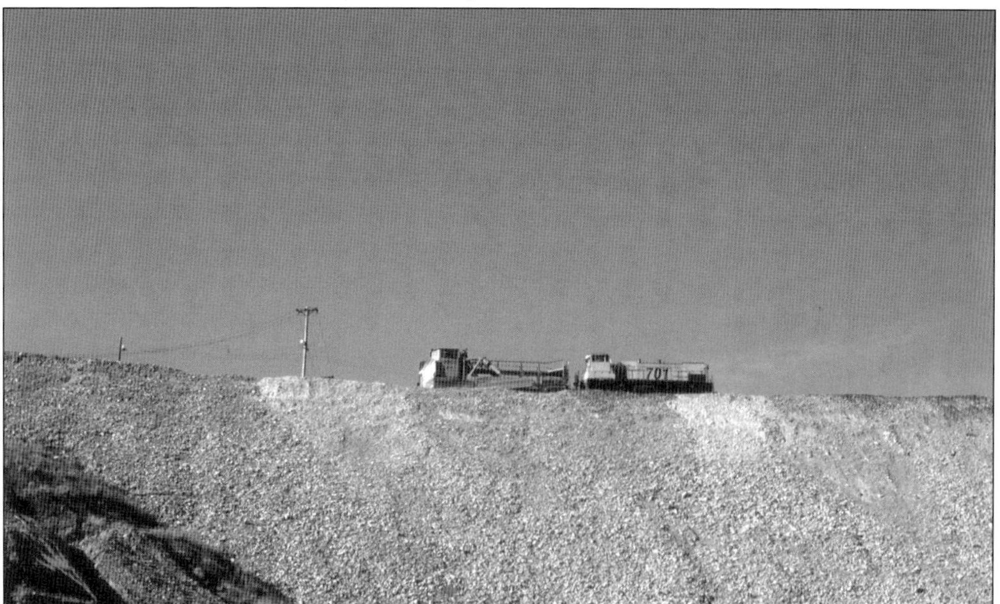

The No. 701 locomotive is at work with the snowplow on the dumps. After a waste train dumped its load of waste rock, a snowplow (Jordan spreader) pushes the material out farther on the dump (away from the rails). The No. 701 locomotive is an EMD model MP15AC diesel-electric with 1,500 horsepower. Kennecott had two of these locomotives—Nos. 701 and 704—and both builder dates were December 1978.

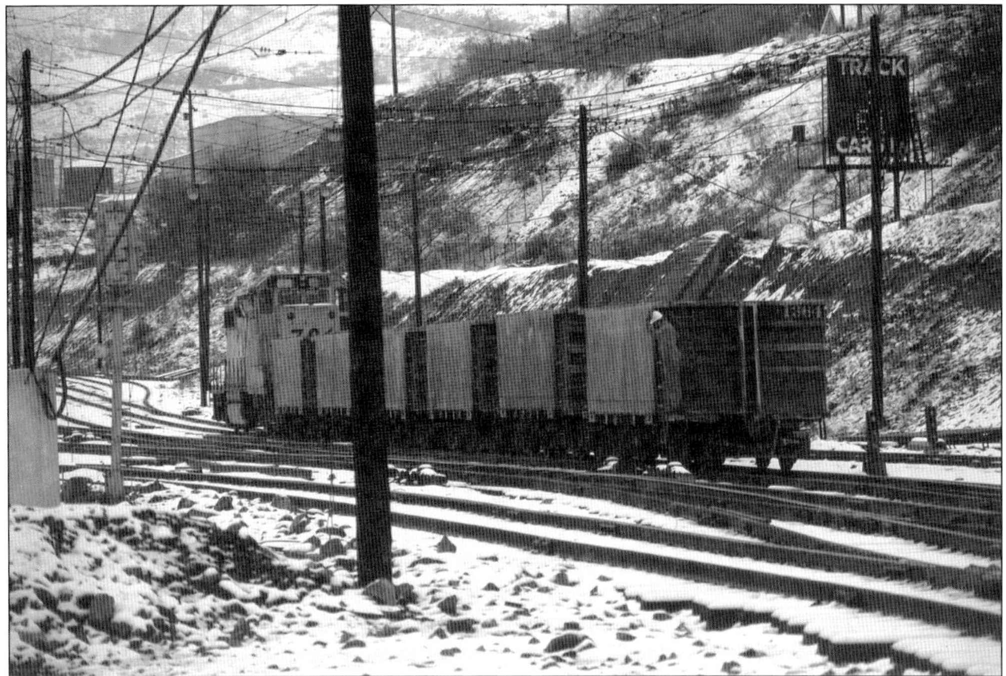

This image shows a diesel locomotive moving six empty ore cars down to Copperton yard, with the brakeman on the end of the train. The brakeman's job was to inspect the track ahead for issues such as broken rails and bad gauge. With the many miles of track in the mine being temporary and most of that track moving constantly, there was always a potential danger.

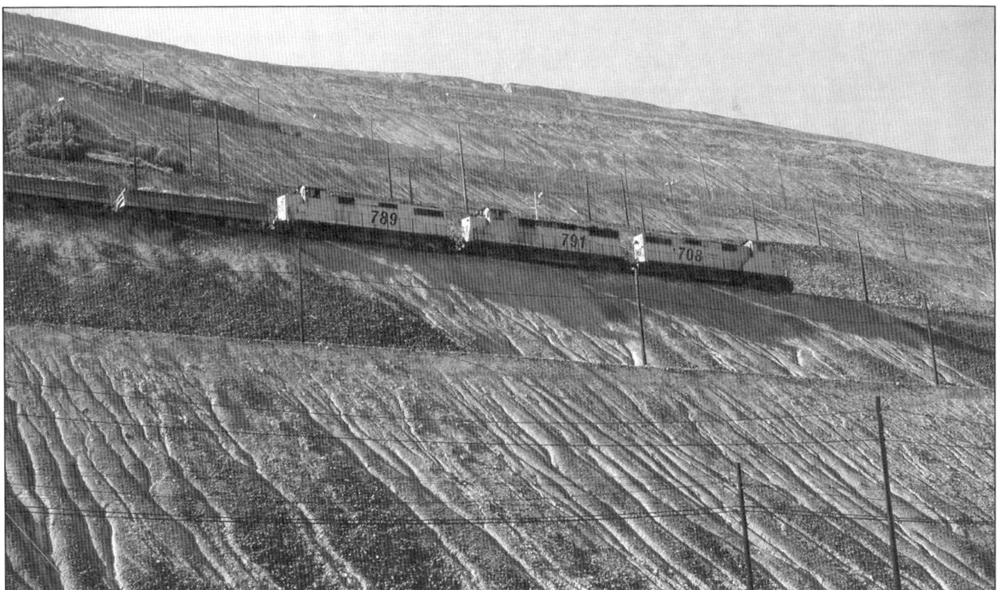

Three EMD GP 39-2s locomotives push a waste train to the dumps. In the 1980s, the railroad operation was just a shadow of what it used to be. Part of the $400 million modernization program, large-haul trucks will move the ore to the crusher and conveyor system, eliminating the railroad. In early 1988, the modernization was completed, but then copper prices soared. Kennecott announced on May 12, 1988, that the railroad would stay as long as the copper prices stayed high.

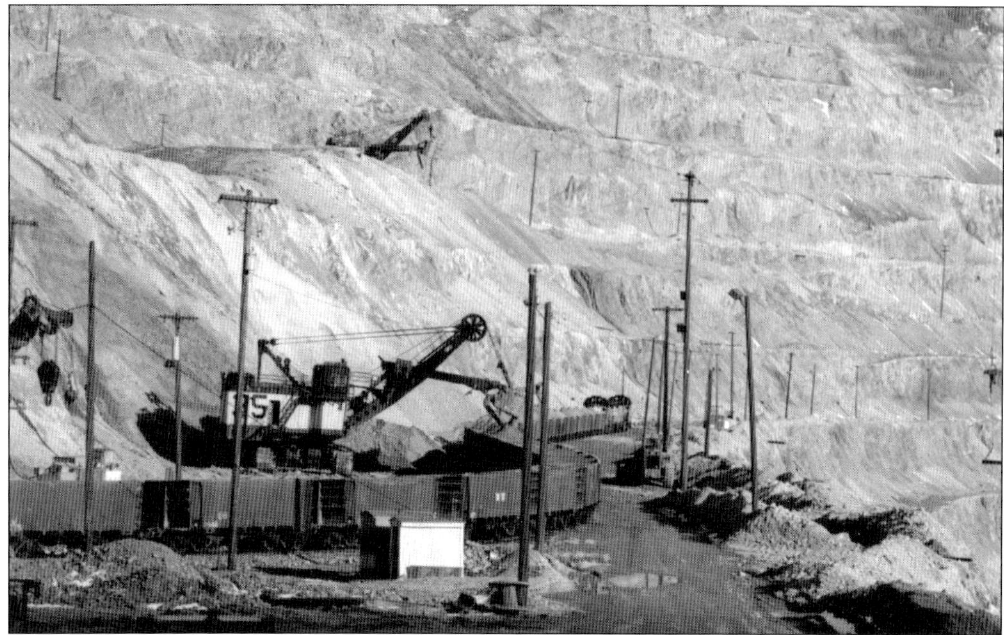

This image shows the last train in the pit being loaded on March 19, 2000. It was announced in 1988 that the trains would have to go, but they lasted another 12 years. From this reload at the 5840 location, the train traveled through the 5840 tunnel down the remaining rail corridor to Copperton yard. The ore was now loaded at the new Dry Fork reload site outside the mine.

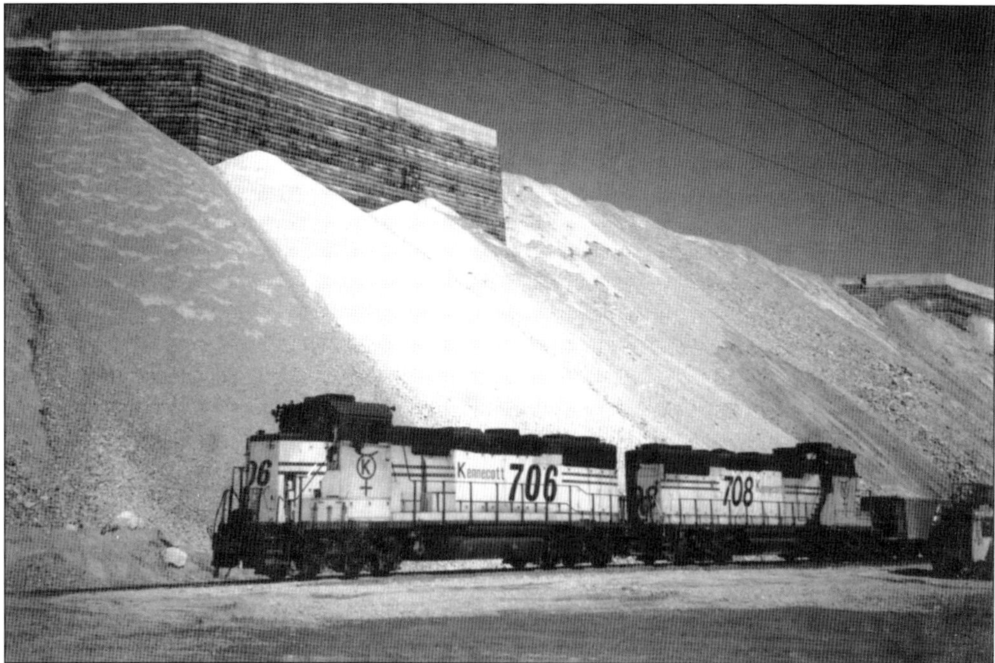

The last train at Bingham was on May 30, 2001. The Nos. 706 and 708 diesel-electric locomotive high cabs are at the Dry Fork reload site. The reload site is below the Dry Fork shops. The previous reload site through the 5840 tunnel in the pit was sitting on top of a pocket of high-grade ore and had to move. This ended all rail services at Bingham Canyon.

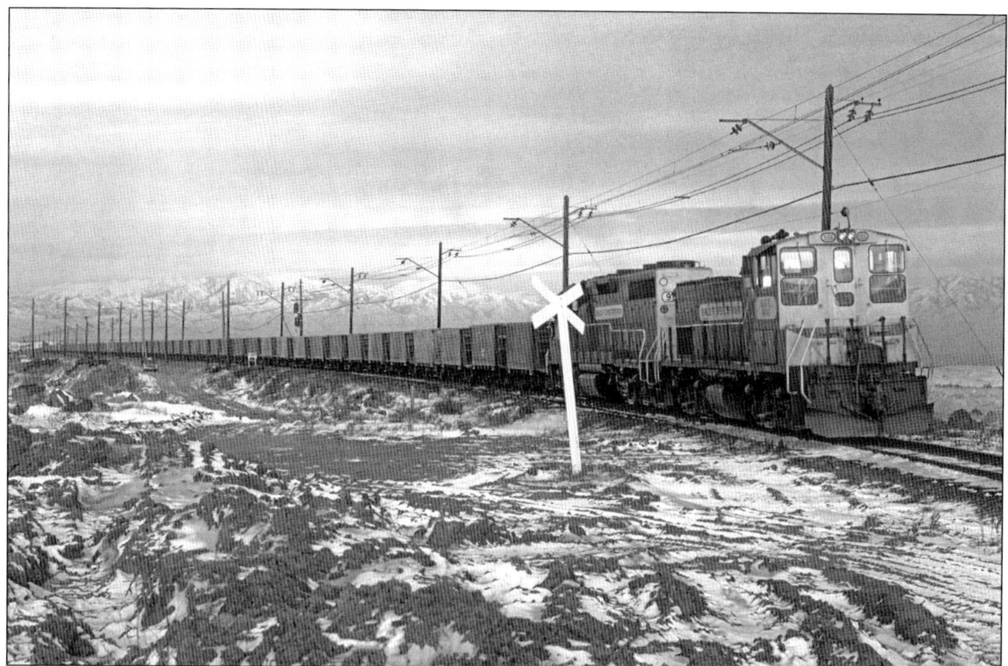

This December 9, 1978, James Belmont photograph shows a long ore train being pulled back to Copperton yard. In the late 1970s, ore haulage was experimenting with diesel-electric locomotives on its main line to the mills. No. 905 was an EMD GP39-2 diesel-electric with 2,300 horsepower. No. 906 was an EMD MP15AC diesel-electric with 1,500 horsepower. Note that the catenary wires are still in place for the large 3,000-volt electrics. (James Belmont.)

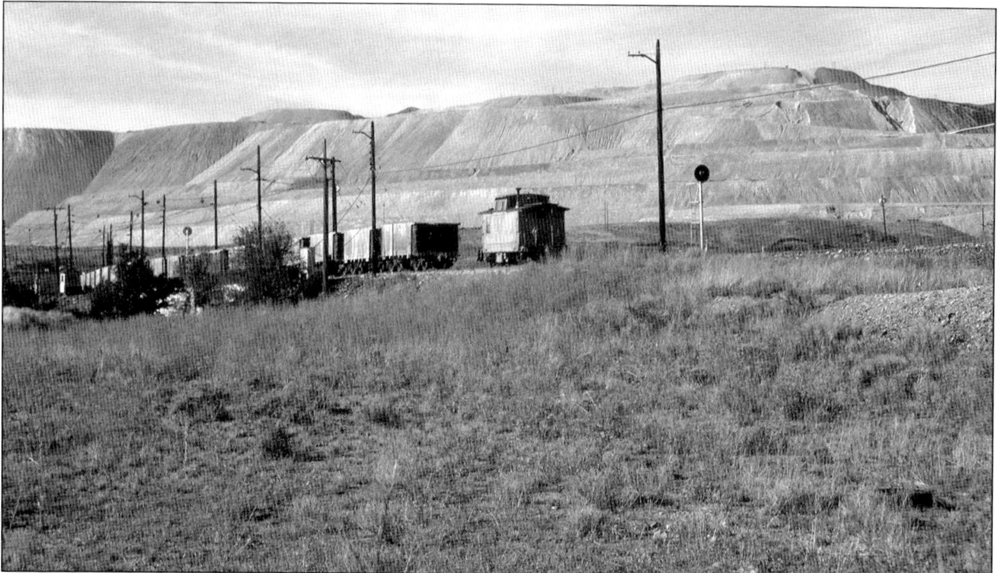

This Don Strack image shows an empty ore train pulling into Copperton yard; the caboose has just been cut off on the fly. In late 1978, Kennecott purchased seven diesel locomotives for its ore-haulage Copperton Low-Line. The locomotives numbered 101 to 107 were EMD SD40-2 with 3,000 horsepower. The new locomotives had a smaller fuel tank of 2,600 gallons because the route was only 16 miles each way.

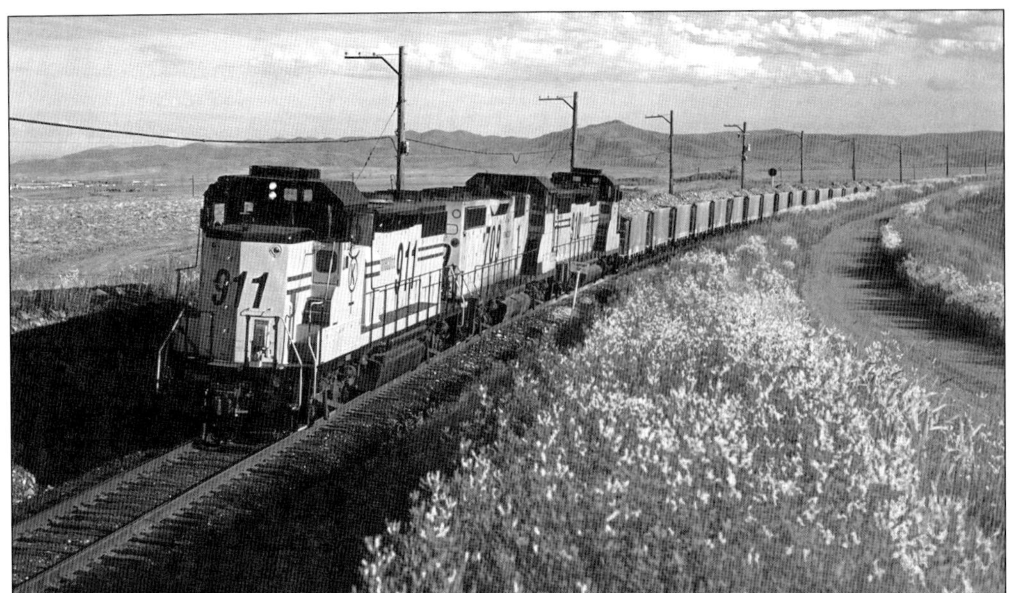

This James Belmont photograph, taken June 11, 1994, shows three of the former pit locomotives on the ore-haulage line. The ore-haulage division was shut down on July 1, 1984. Train crews from the mine now moved the ore to the mills. The high cost of copper in the 1980s and 1990s kept the trains running. The railroad now moved 35,000 tons a day with 400 ore cars. (James Belmont.)

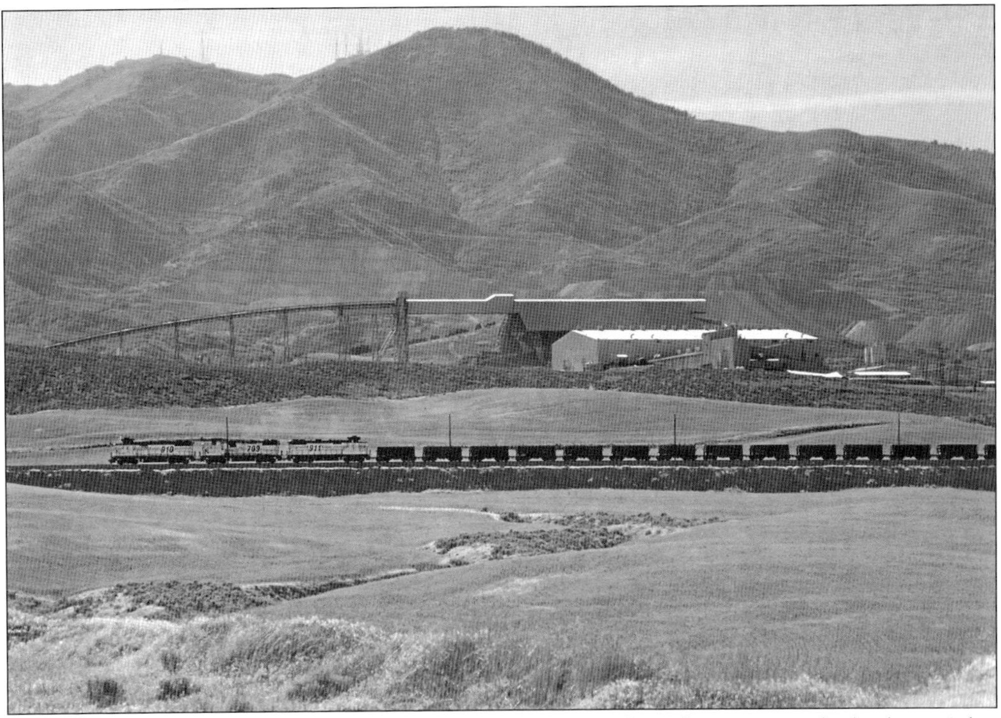
This historical image shows an empty ore train heading back to the mine. In the background is the new Copperton Mill, which eventually replaced the railroad. The new concentrator complex produced 77,000 tons of concentrate a day, with a lot fewer employees. The frothy mineral mixture was pumped 14.02 miles in a six-inch pipeline to the smelter, replacing the train. (James Belmont.)

Nine

The Work

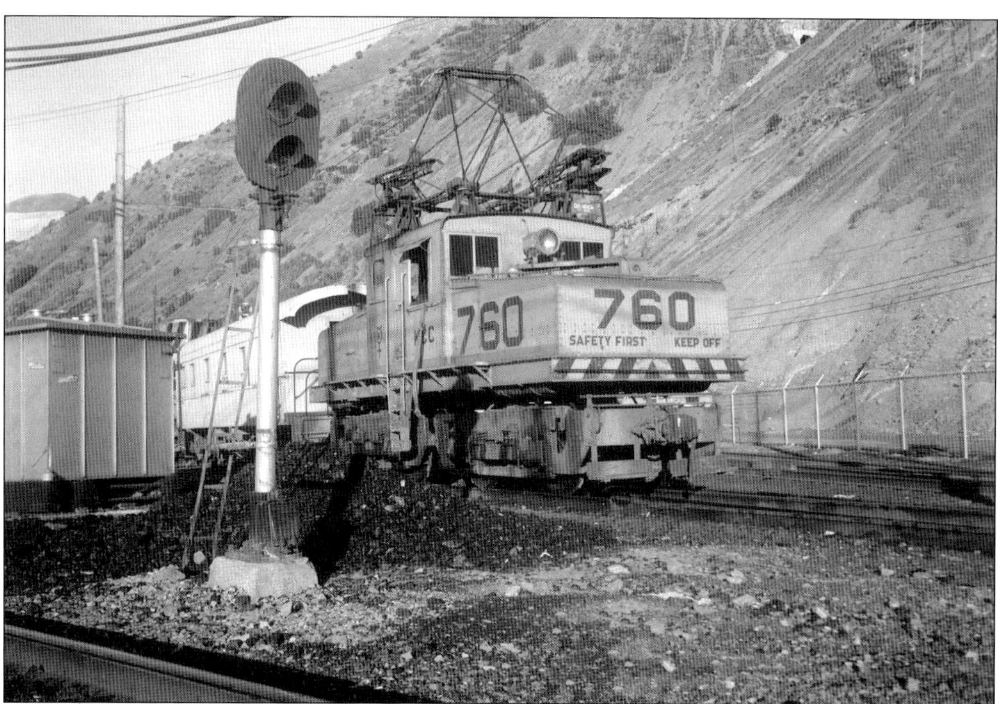

This image shows the No. 760 electric locomotive pulling the work coach. Many employees entered the mine by coach, train, and shovel crews, and track gangs traveled to their job location in the mine. Open-cut mining is about getting the ore out of the ground of the mountain and moving it to the mills for processing with the most cost-effective method possible. Kennecott has been doing that for over 100 years.

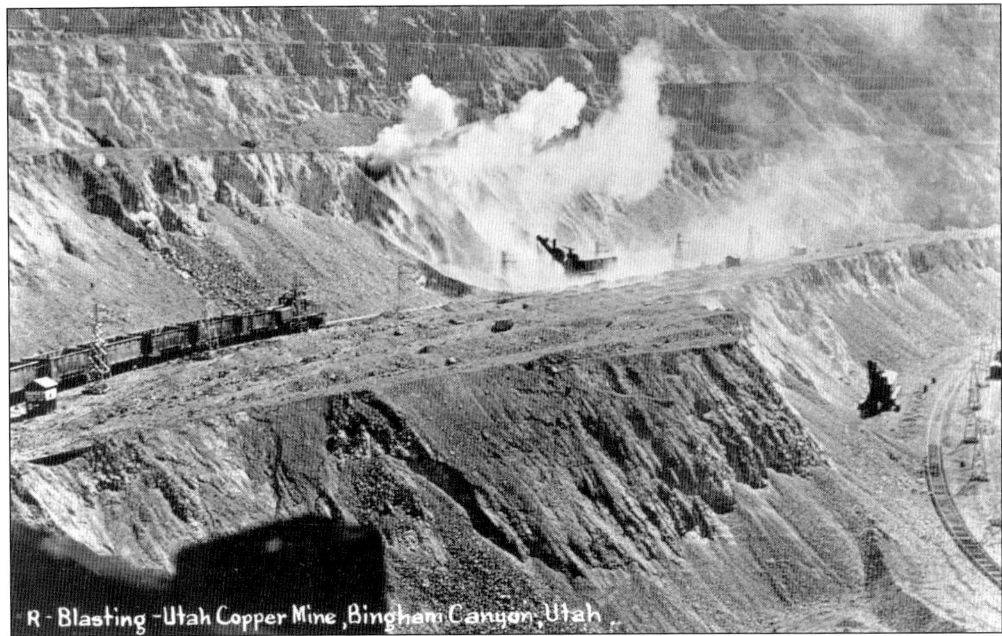

The train waits as the drill and blast department detonates another blast—an everyday occurrence at the Bingham Canyon Mine. Hard rock mining takes a lot of explosives; a 1974 fact sheet stated that 85,300 pounds of explosives and blasting agents were used per day. In 2024, approximately 200,000 pounds were used per day. For every pound of explosive used, 4.11 tons of material can be mined.

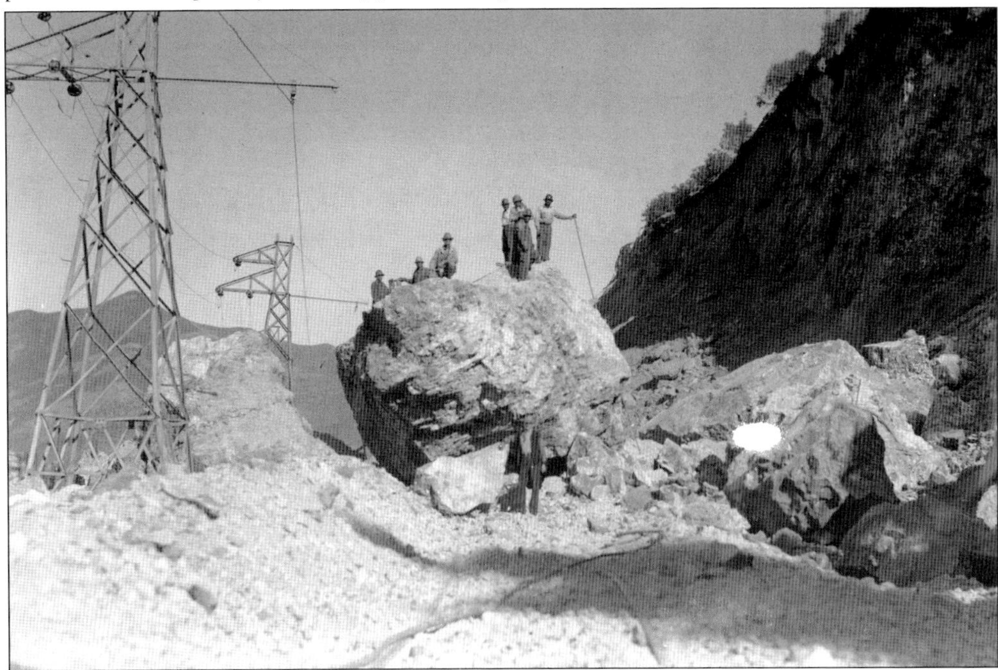

This image shows a large boulder on the railroad tracks, and members of the drill and blast department pose for a photograph on top of it. To remove the large boulders, the drill and blast department drills holes in the rock and then fills the holes with dynamite to break the rock apart to a workable size. Hercules Powder Company supplied the dynamite; it came to the mine in boxes.

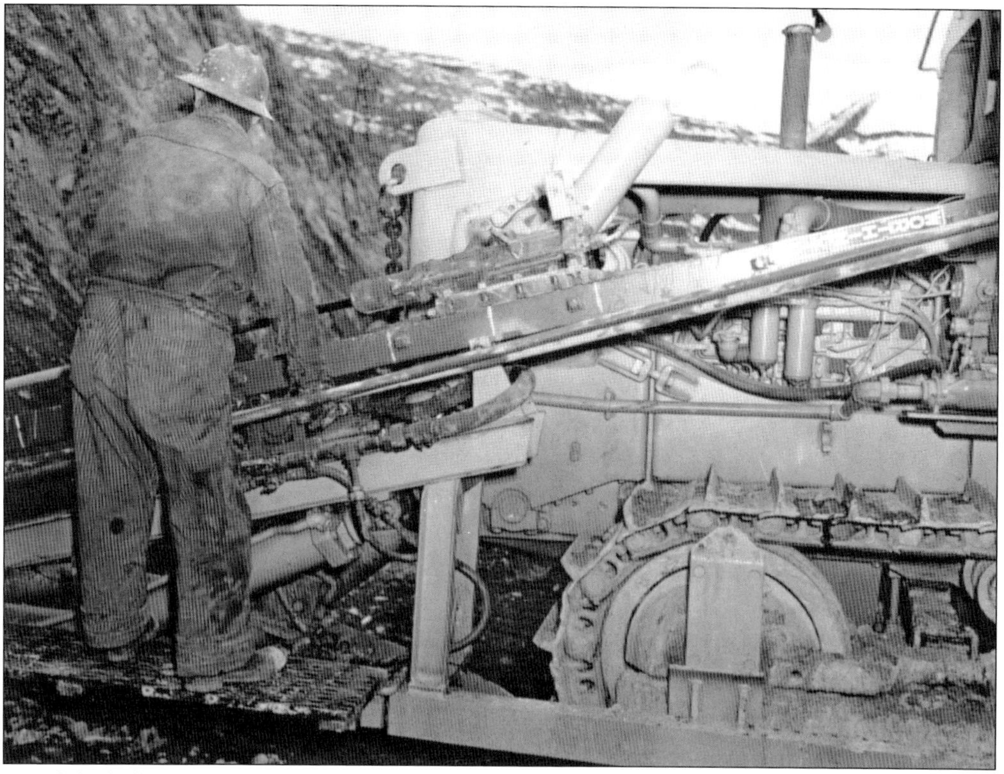

A mobile drill bores into the side of the level. This was once done by a pneumatic drill on a tripod. This method is called toe drilling, and blasting techniques have changed many times over the years. Once completed, samples are taken every 10 feet, and those samples are assayed to determine the percentage of copper. The results are placed on a plan map, and the survey crew stake off the level where the waste ends and ore begins.

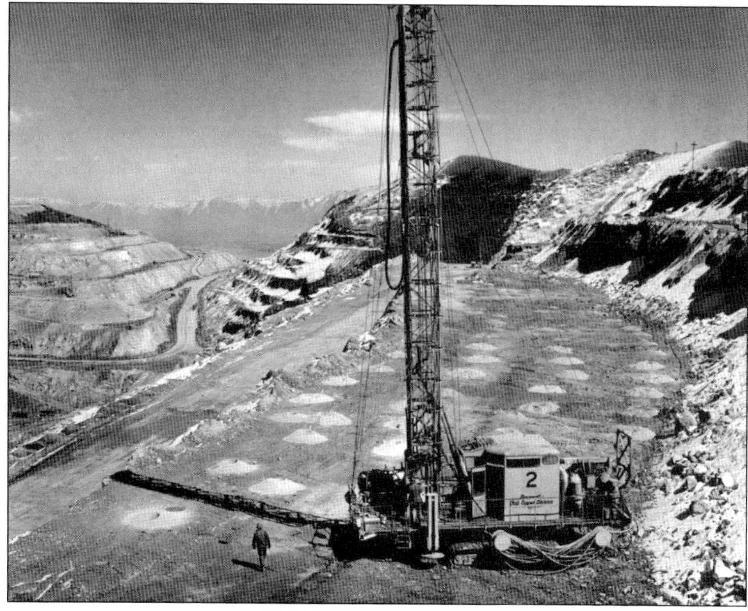

This image shows the No. 2 rotary drill in the upper portion of the mine. The many holes that are visible will be filled with a blasting slurry solution and then detonated. Once the large rotary drills came to the mine, it became the preferred method. In 1974, Kennecott had 18 rotary drills: nine Bucyrus-Erie 60R 12.25, eight Bucyrus-Erie 30-R 6.75, and an Ingersol-Rand DN-4 6.75.

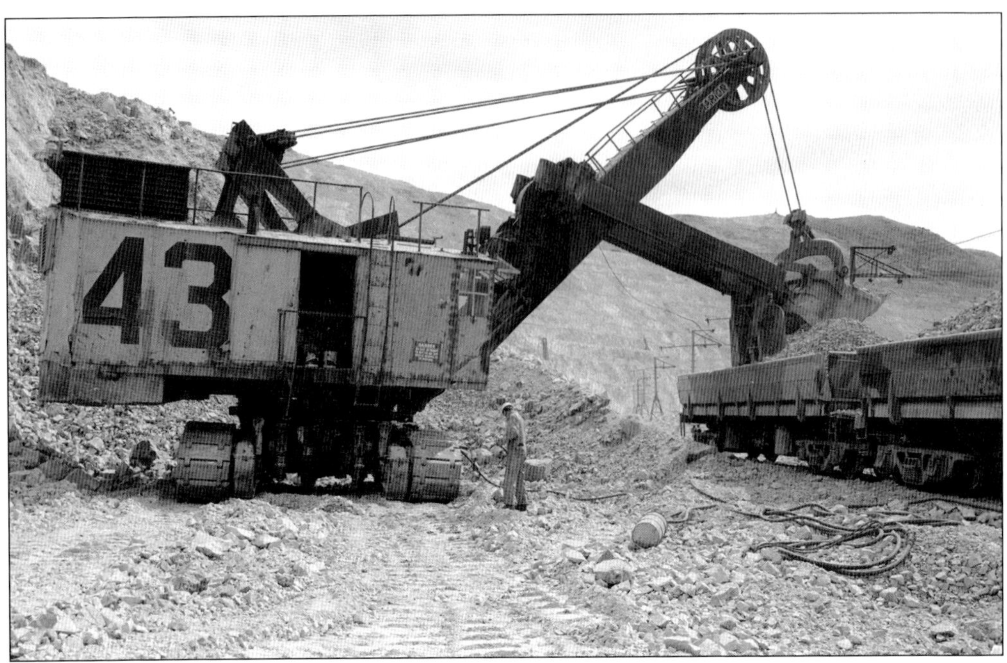

Electric shovel No. 43 is seen here loading a waste train. Shovel crews were made up of two men, the operator and the oiler, and both took turns running the shovel. Shovel and train crews worked three shifts a day around the clock. Shovel operators were some of the highest-paid employees at the mine. The man on the ground is probably the oiler; he is adjusting the trail cable that provides power to the shovel.

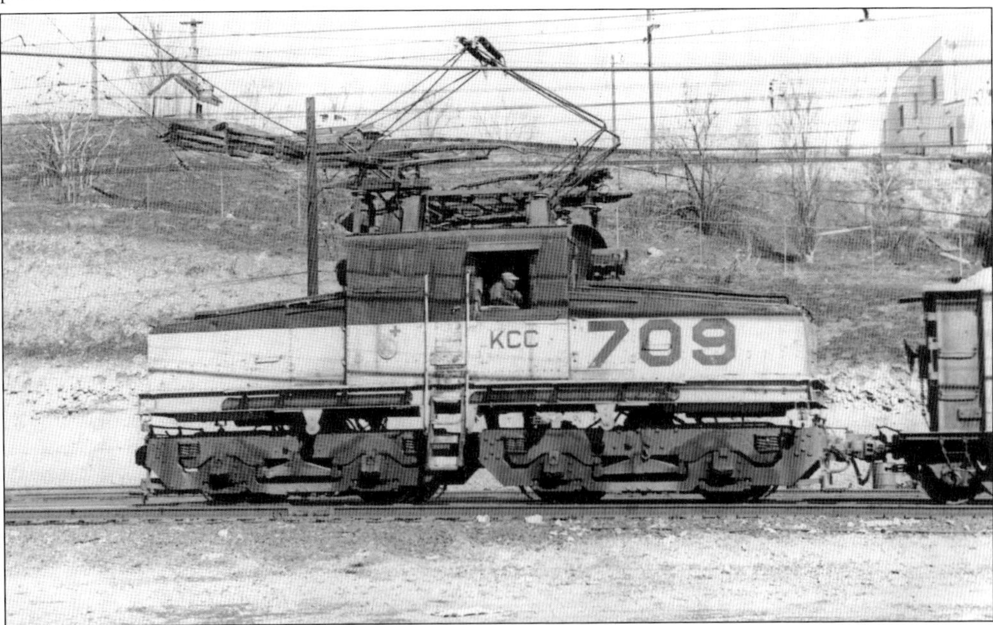

The No. 709 locomotive pushes loaded ore cars into Copperton yard. Train crews were made up of two-man crews: the engineer and the brakeman. At Kennecott, the engineer was called a "hogger," an old railroad term. In the railroad heyday, Kennecott ran 22 train crews. To bring a 2,700-ton train of rock down a four-percent grade took skill, thought, and a lot of attention.

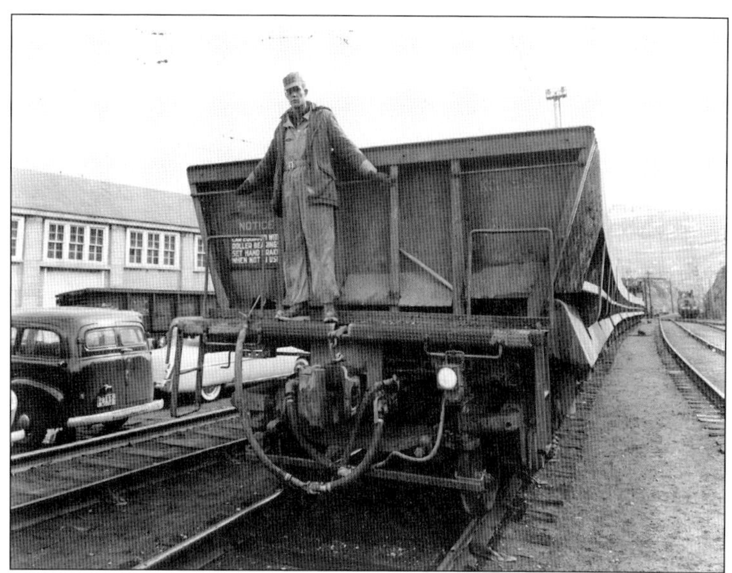

This image shows a brakeman on the end of a waste train as it moves through the B&G yard. For trains coming out of the mine traveling car first, the brakeman was required to ride in front of the trains; his job was to keep eyes on the trackage ahead for problems. If there was trouble, the brakeman could step on the "tail-hose," a device connected to the train's air line to make an emergency stop, shown center left.

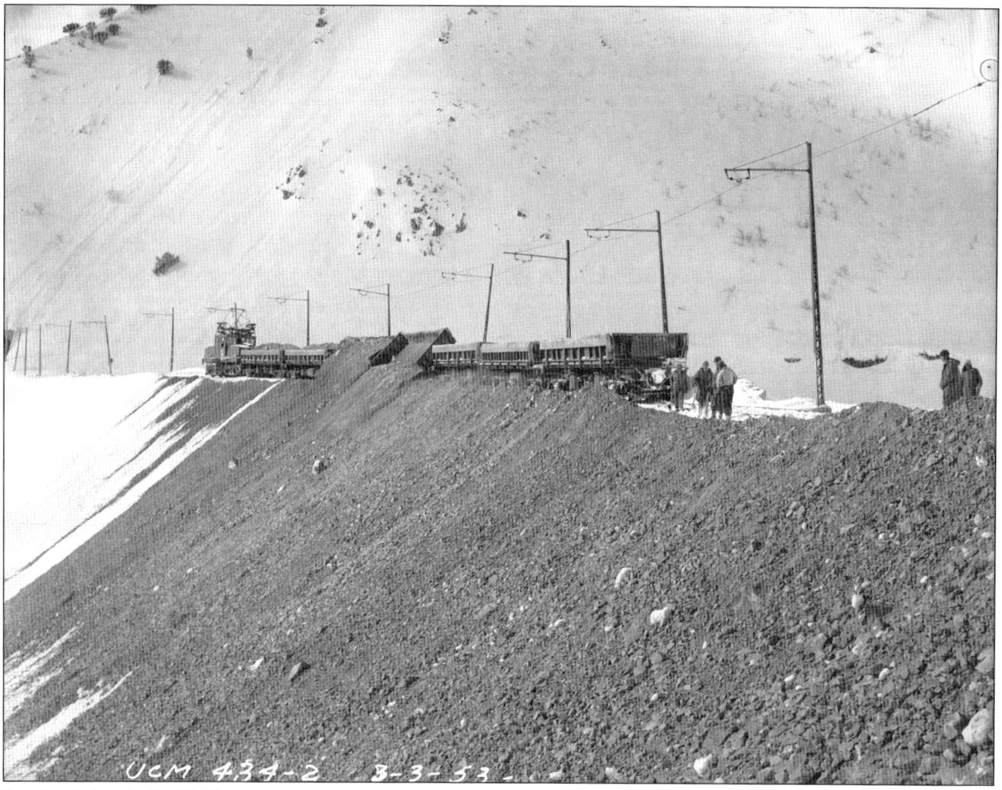

In this March 3, 1953, image, a waste train is dumping its load. The track gang is at the end of the train. The image shows the importance of a snowplow to push the overspill over the dump. The dump track gang's job was called "shaking out." Working with a track shifter, the shifter would pick up the track, and the men would fill in under the railroad ties, fix downed ties, and resurface the track.

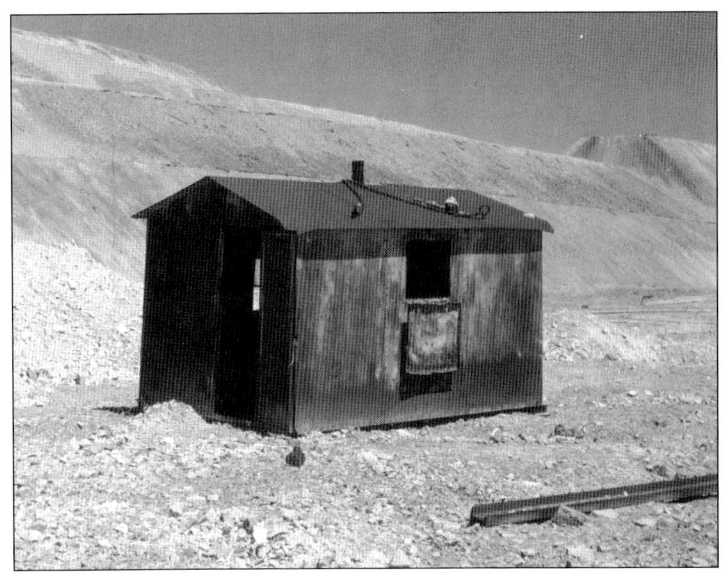

This image shows the track gang crew shack out on the dumps. The shack usually stayed in one place, and in the mine, the track gang shacks traveled to wherever the gangs moved to, thus the chain and eyehooks on the roof. A crane (or jitney) would pick up the shacks, put them on a flat car, and move them to the job site. The shacks were the only protection from the weather that the gangs had.

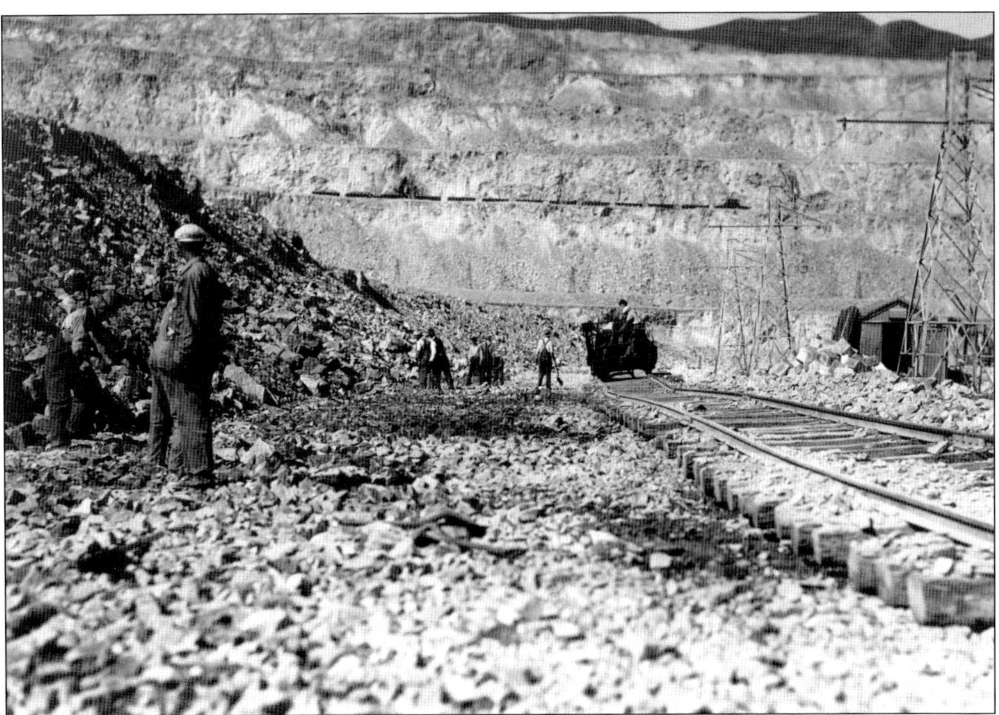

The track gang and a track shifter are pictured working in the mine, also called a "line job." After the shovel made a pass loading the train cars, the track had to be moved closer to the level. The shifter would clamp onto the rails, and then a spud would come down between the ties, lift the track up, and shift it over.

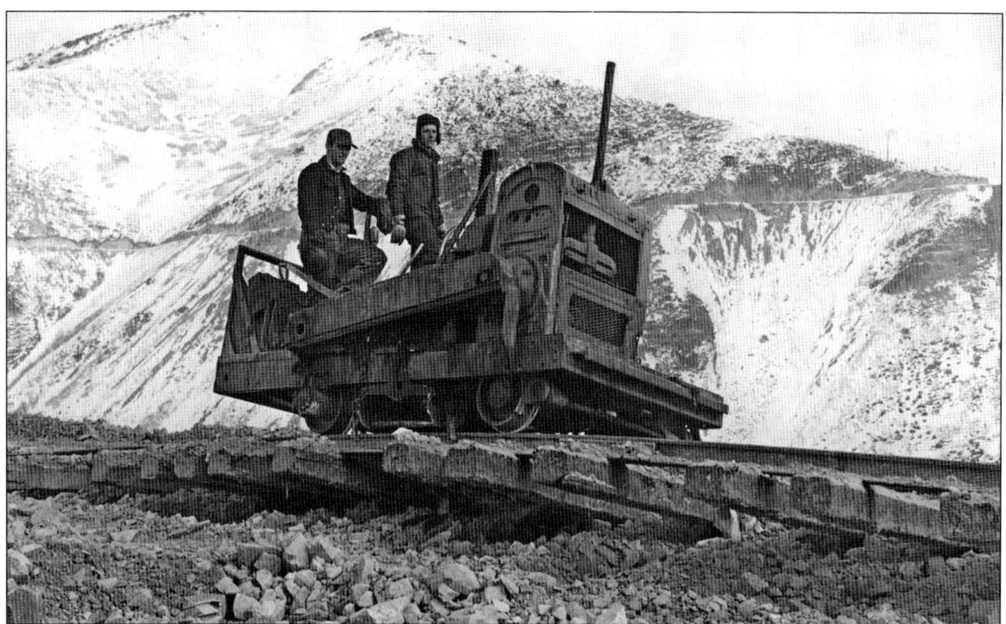

This is a close-up view of the track shifter. This was a two-man job, requiring an operator and a man to throw the clamps. The track shifter is lifting the track out of the overspill, and now the operator can shift the tracks over. Two shifters worked together on line jobs. A large section of track would be broken apart at both ends, and once the track was moved over, it was pieced back together.

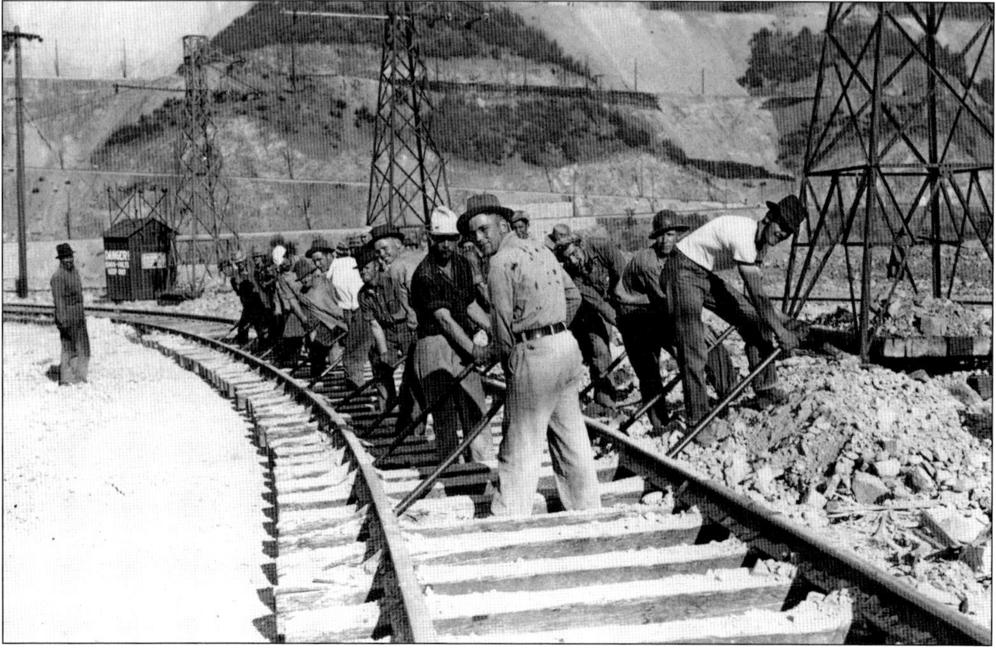

This image shows the track gang shifting the track the old way. The gang probably did not have to move the track very far. The foreman would stand back and eyeball the track to see how far it should be moved—this was called "lining the track." The track gang was the entry-level job at the mine; men started here and then could bid out to different departments when they accumulated enough seniority.

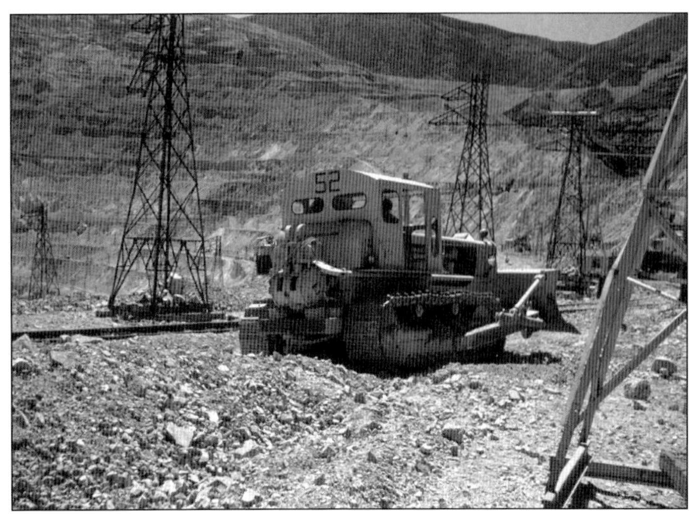

A dozer is seen here working in the mine. The dozers and graders had many different jobs around the mine. Dozers leveled off the area for the line job and helped move the track over and push waste rock over the dumps. At truck haulage, rubber tire dozers kept the shovel overspill cleaned up where the haul trucks were loaded, and dozers built safety berm along the haul roads. Graders kept the roads smooth and rock-free to save on tire costs.

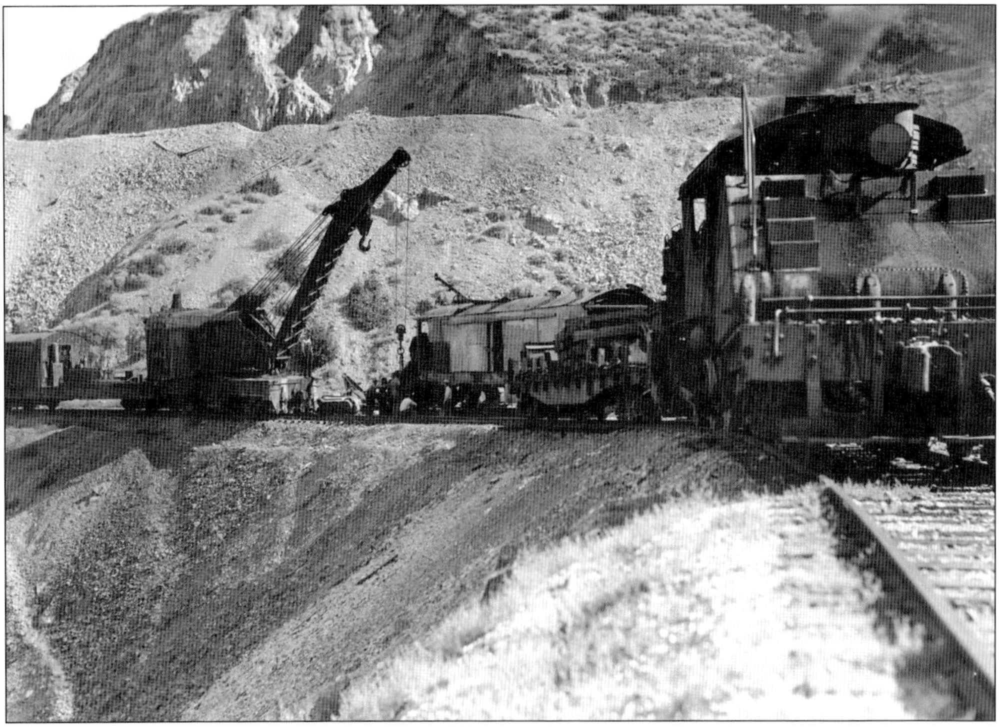

This bull gang works on an old standard shovel just outside of the mine. The bull gang personnel were heavy-duty mechanics and welders and did whatever the job took. The bull gang was literally a dirty job; the electric shovels with all the gears and moving parts needed a lot of lubrication, and coveralls were a must. Shovel crews had the operator and the oiler; the oiler greased and oiled the shovel.

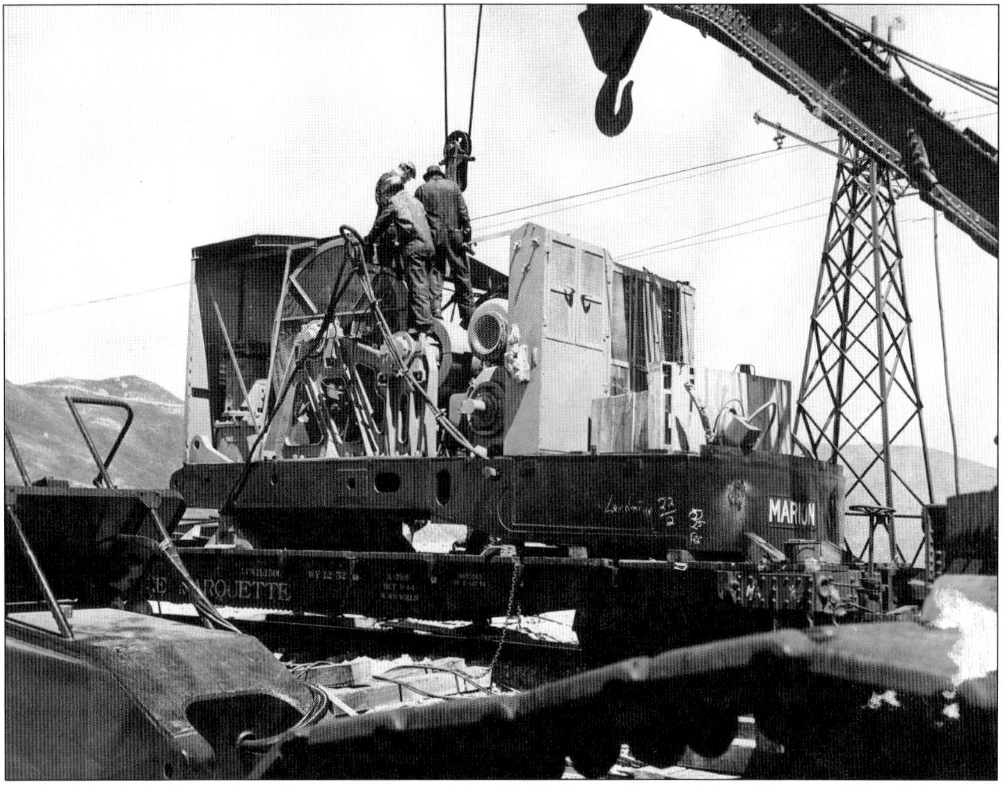

The bull gang is assembling a new Marion electric shovel at the mine. The shovels were built at the Marion factory, disassembled and delivered to the mine on railcars, and then reassembled at a wide spot on the level. This electric rope shovel was called a whirling shovel because the whole shovel moved. It had a six-yard dipper and could load an ore car in five minutes.

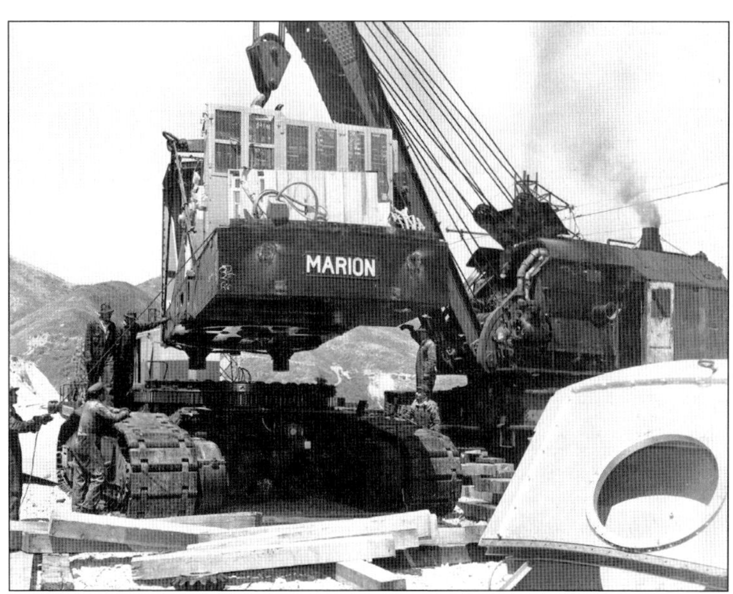

This image shows the big hook (or derrick crane) lifting the main body of a new electric shovel on its frame. The crane department at the mine ran the large derrick and small jitney cranes that ran on rails and some mobile cranes. The big hook worked on the shovels or on train wrecks, and the jitney helped lay new rail, move the portable towers, and perform other odd jobs.

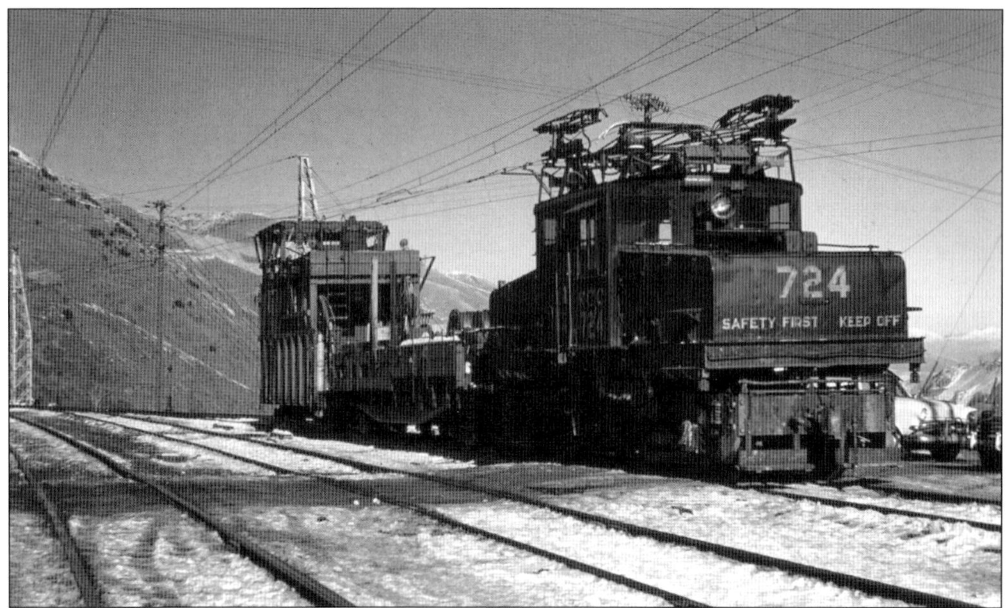

The No. 724 electric locomotive and the line car sit in the B&G yard in 1957. The line car was specially designed to work on the overhead catenary system at Bingham. With over 100 miles of electrified railroad, much of it temporary and on the move in the mine, it took a lot of work to keep it up and running—this was the job of the electricians at the mine.

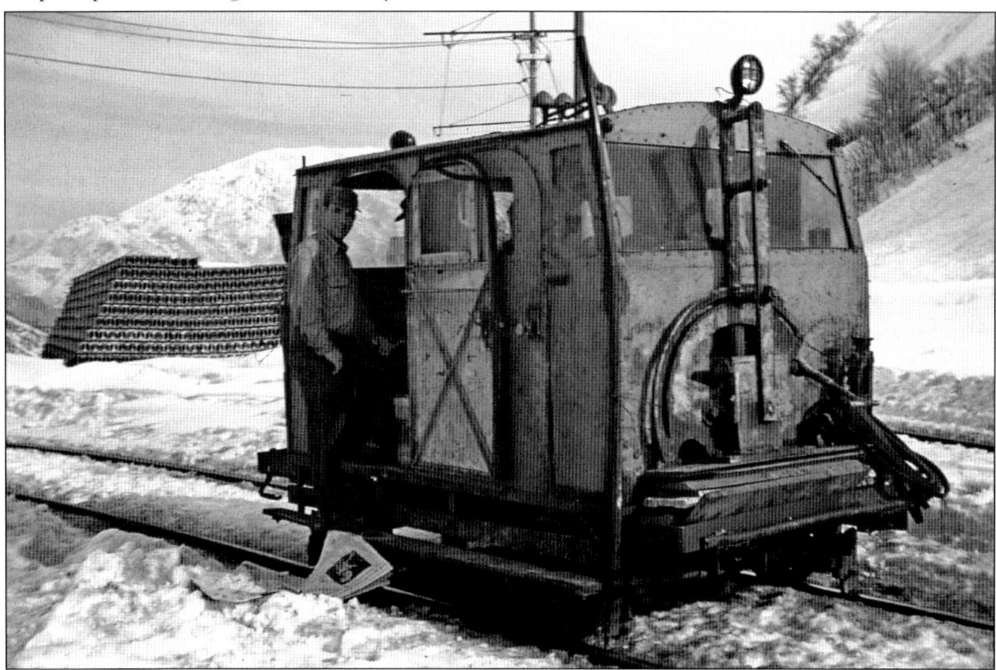

The image shows a speeder and operator in the background, with stacks of new rail. The speeder had many jobs around the mine, but its main job was to weld copper connections between the rails (known as "rail bonding"). The speeder operators also worked on the rail switches. By 1974, the mine had 106 rail switches, and most were electrified. The speeders had tools to heat and bend the rail where the switch points went.

A switch tender shack at the mine is pictured here. Before the mine had electric switches and centralized traffic control (CTC) and two-way radios in the locomotives, train movement in and out of the levels and up and down the switchbacks was controlled by switch tenders. Some locations had multiple switches the switch tenders had to keep track of. In the wintertime, the switch tenders had to keep the switches clean and free of snow.

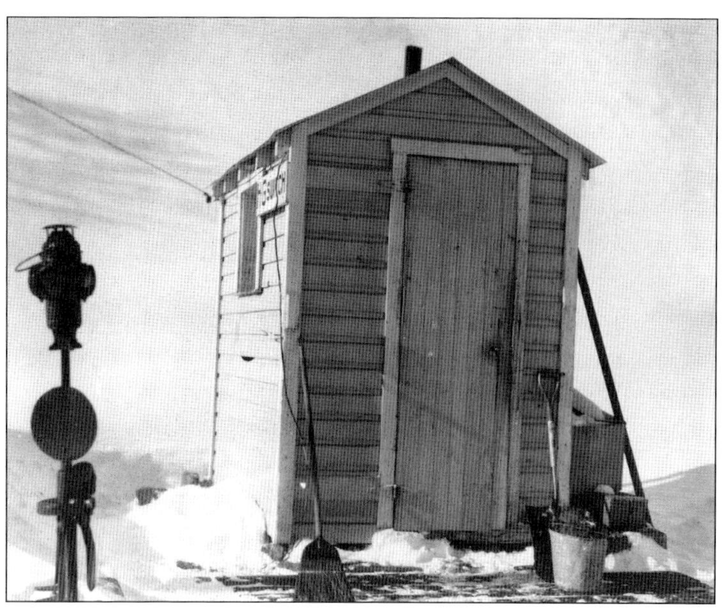

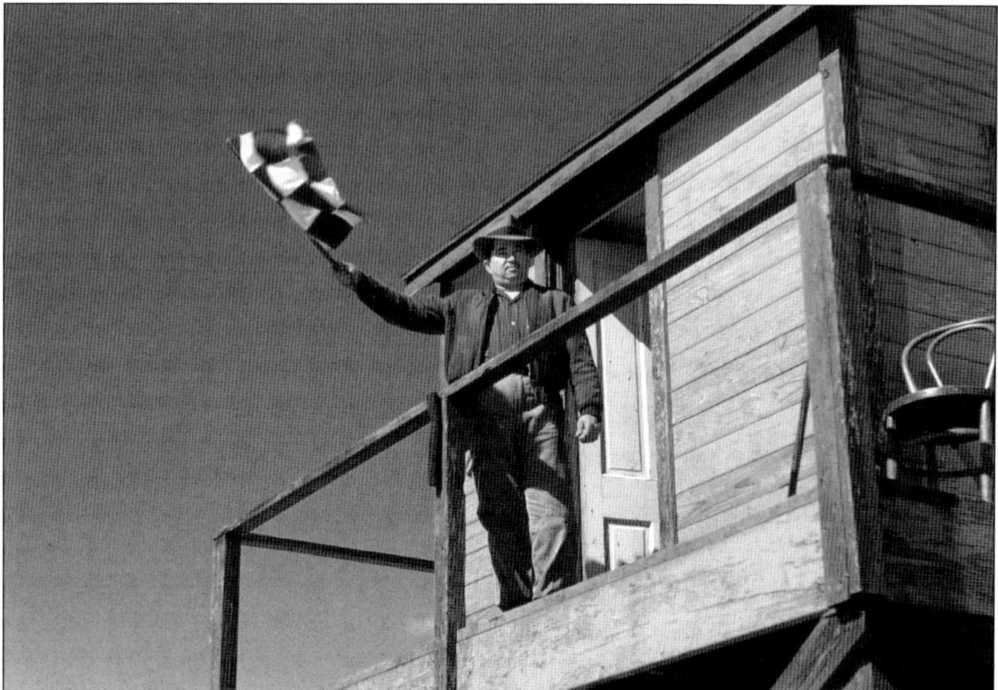

This image shows a flagman waving a train ahead. Before centralized traffic control and two-way radios, the mine had flagmen and flagman towers to navigate trains around sharp corners and blind spots at the mine. Some of these flagman towers were built very high to be able to signal distant trains. The flagman and switch tender jobs at the mine could be lonely, with men being alone for eight hours at a time.

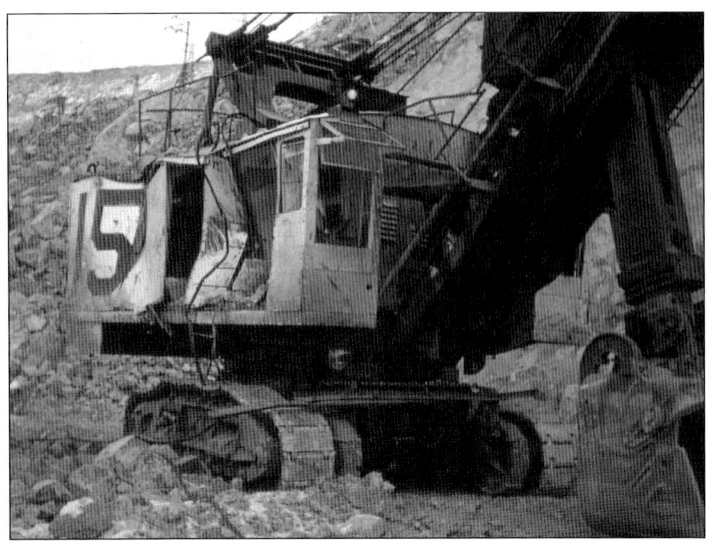

The No. 15 electric shovel is seen here smashed in due to a landslide. Landslides, wrecks, and accidents occurred at the mine. Having over 7,000 Kennecott employees and the many moving parts to the mining operations, the company's safety record was good. The work at the mine was around the clock and performed in every kind of weather. Working in the middle of the night or on a rainy or snowy day had its challenges.

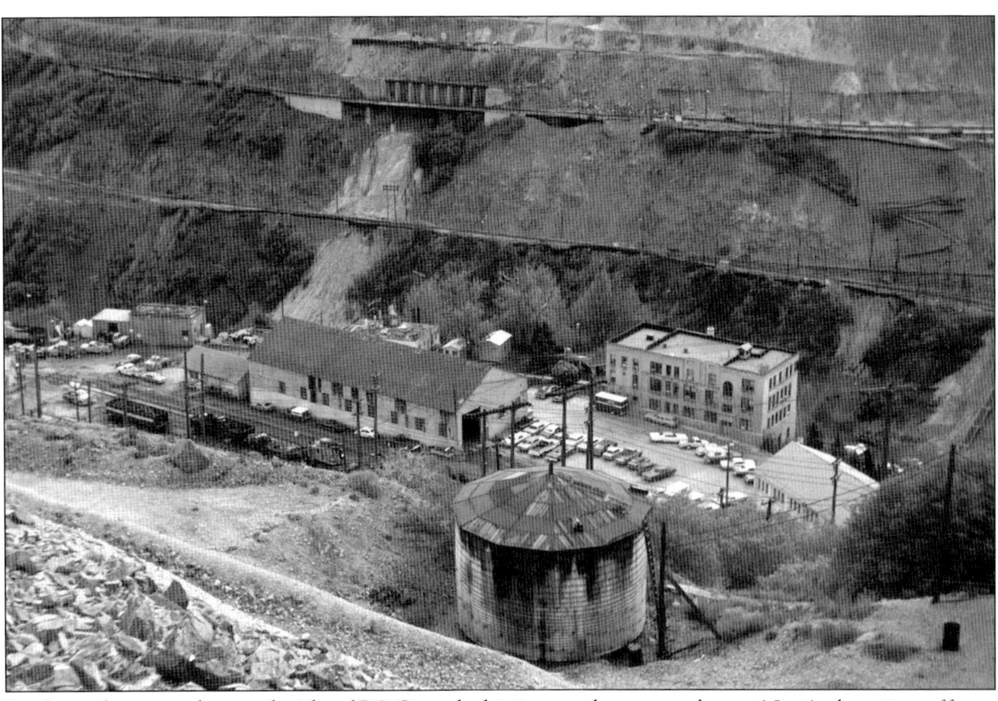

Looking down on the north side of B&G yard, this image shows warehouse No. 1, the mine offices, assay department, the oil house, electric field office, and the material shed. Moving the ore from the mine to the mills also took a long list of support people doing all the work behind the scenes. Mechanics, welders, service truck drivers, water service, machinists, and payroll—there are many important roles to mention.

Ten

Mine Expansion

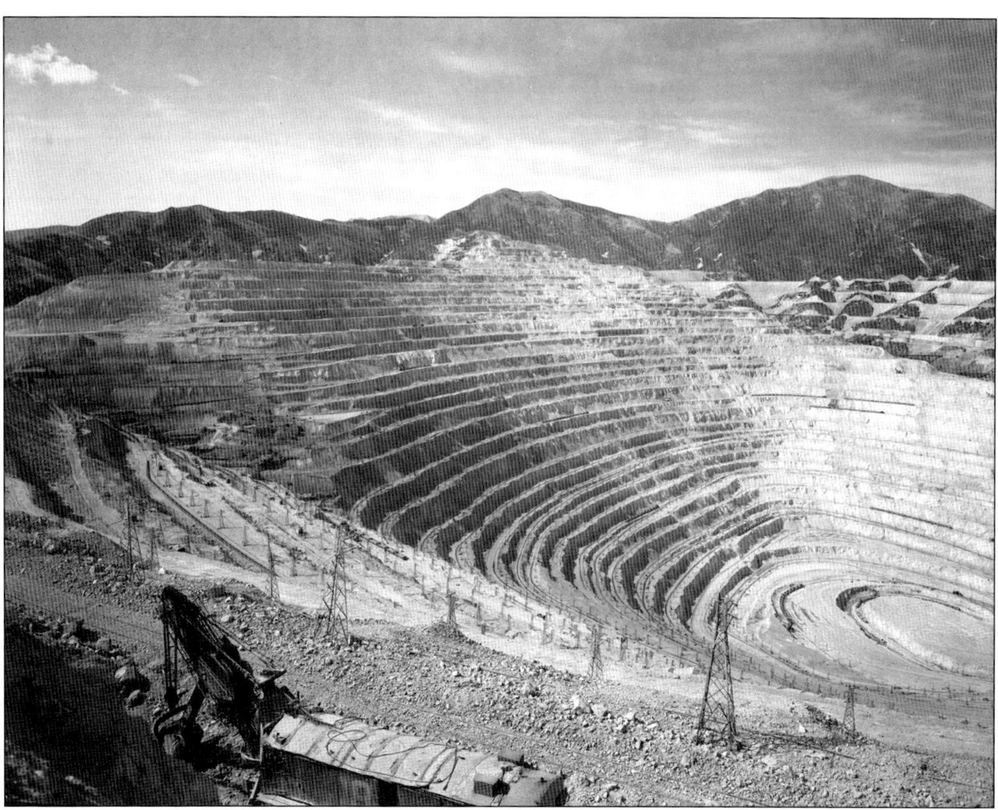

This view looks northwest over the expanse of the mine in the 1950s. From the beginning of open-cut mining, the mine was always changing and improving. Daniel Jackling's idea to mass-produce low-grade copper on a large scale was paying off. Utah Copper made profits from the beginning, and much of that profit went back into expanding and upgrading the mine, mill, and smelter. Jackling was always looking for a bigger and better way of doing things.

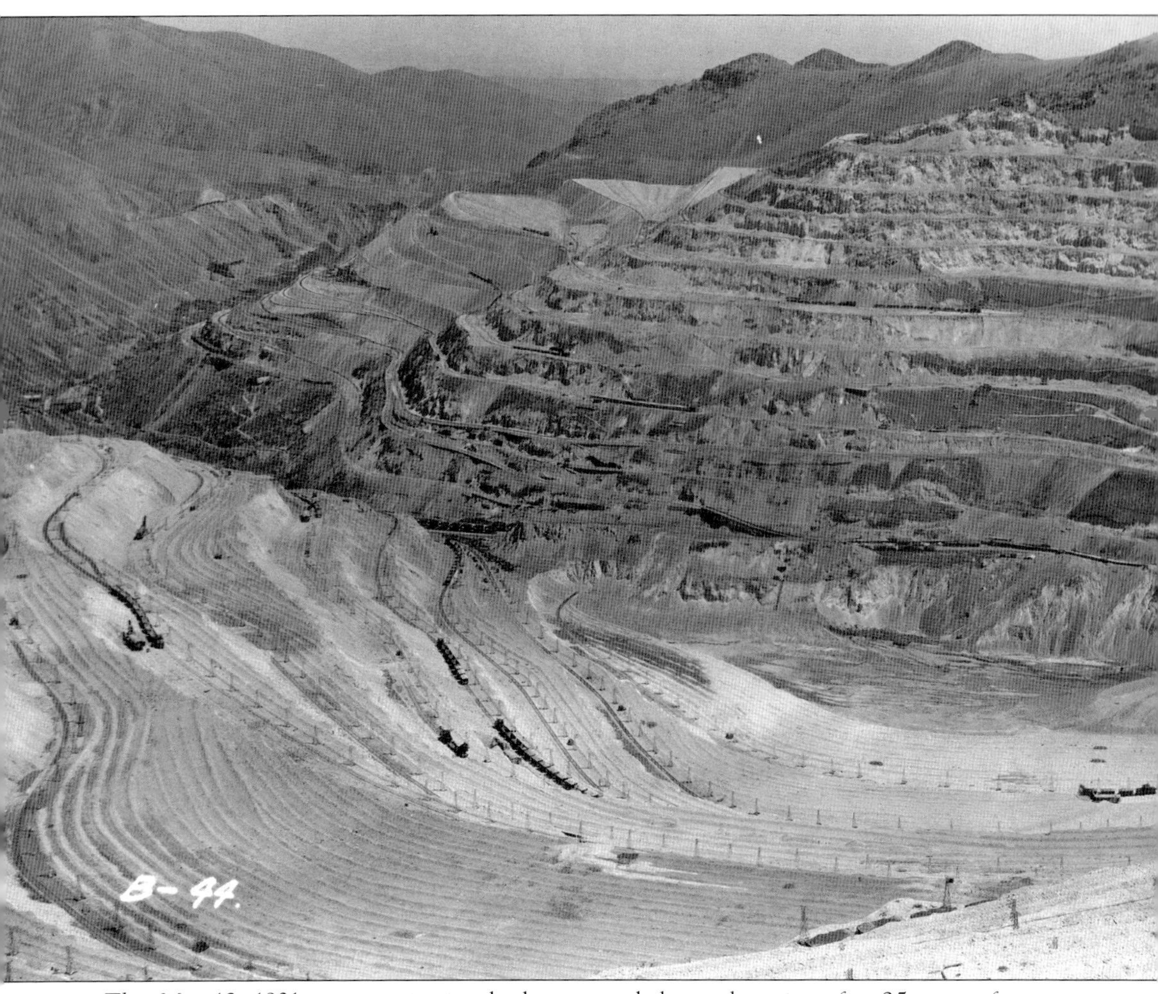

This May 12, 1931, panoramic view looks east and shows the mine after 25 years of open-cut mining. Much of the mountain that was once there has been eaten away. The mine is being dug deeper, going below A-level (6,340 elevation), making a pit. Once the mine went below A-level, the levels were named by their elevation instead of by letters—shown are 6340, 6290, and 6240, the bottom of the pit. By the time this photograph was taken, many mining methods had changed.

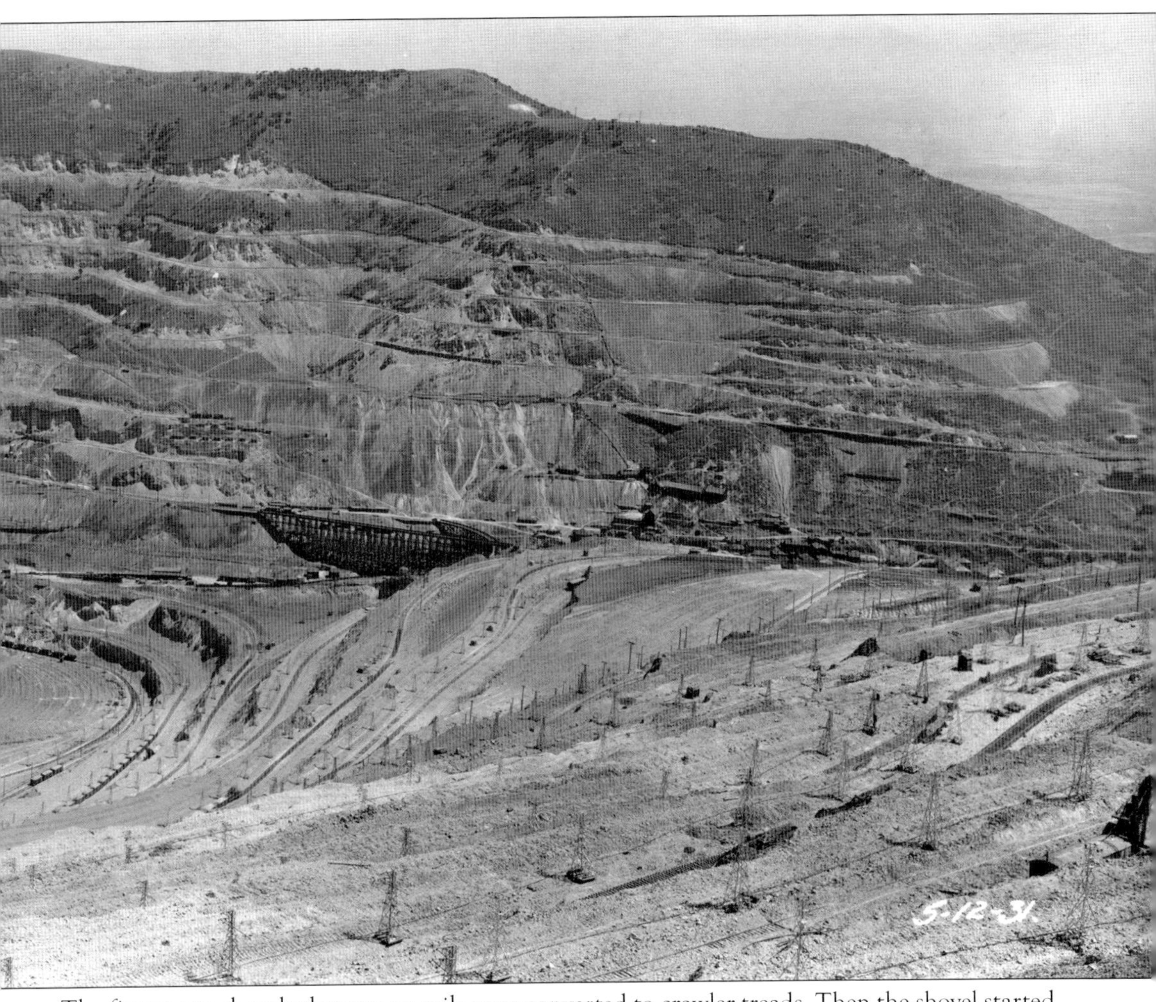

The first steam shovels that ran on rails were converted to crawler treads. Then the shovel started running on electricity instead of steam, followed by the steam locomotives switching to electricity. The mills and smelter were also going through changes. The Garfield smelter furnaces first used coal and then oil in 1911. The year 1915 brought powdered coal, and natural gas came in 1930.

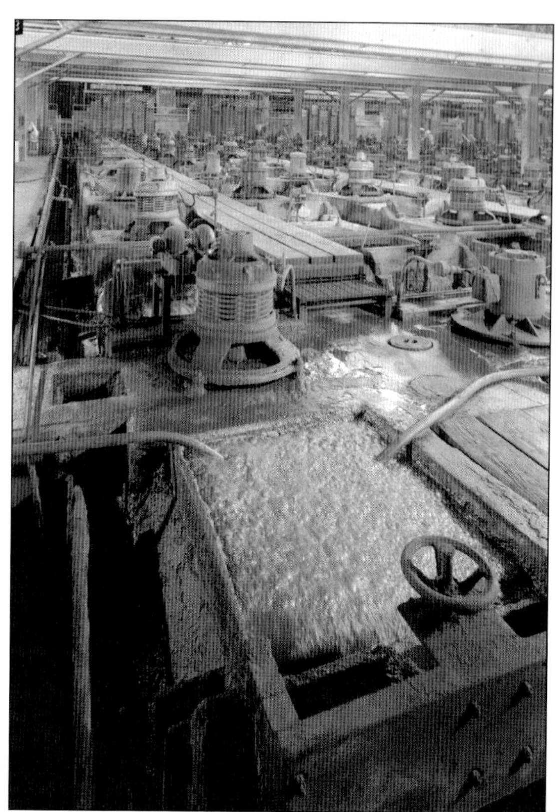

This image shows the flotation process at one of the mills. There were two mills—the Boston Consolidated Mining Company and Utah Copper—and these two companies merged in 1910, with the Boston Consolidated mill name changing to the Arthur mill and the Utah Copper mill becoming the Magna mill. In 1922, both the Arthur and Magna mills made extensive improvements in froth flotation, increasing the percentage of ore from 73 percent to 81 percent.

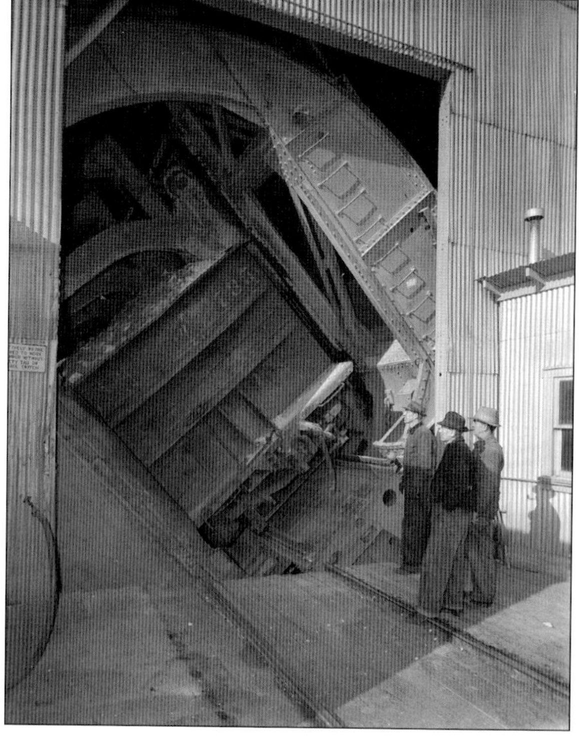

Seen here is a Magna car dumper in 1942. One of the most cost-effective improvements at the mills was adding the car dumpers. Wellman-Seaver-Morgan installed a single car dumper at the Arthur mill; it went online on January 15, 1919. In 1923, the Magna mill installed a twin car dumper. The car dumpers greatly increased the amount of ore moving into the mills. A Magna new twin car dumper was capable of dumping 720 cars a day (48,000 tons).

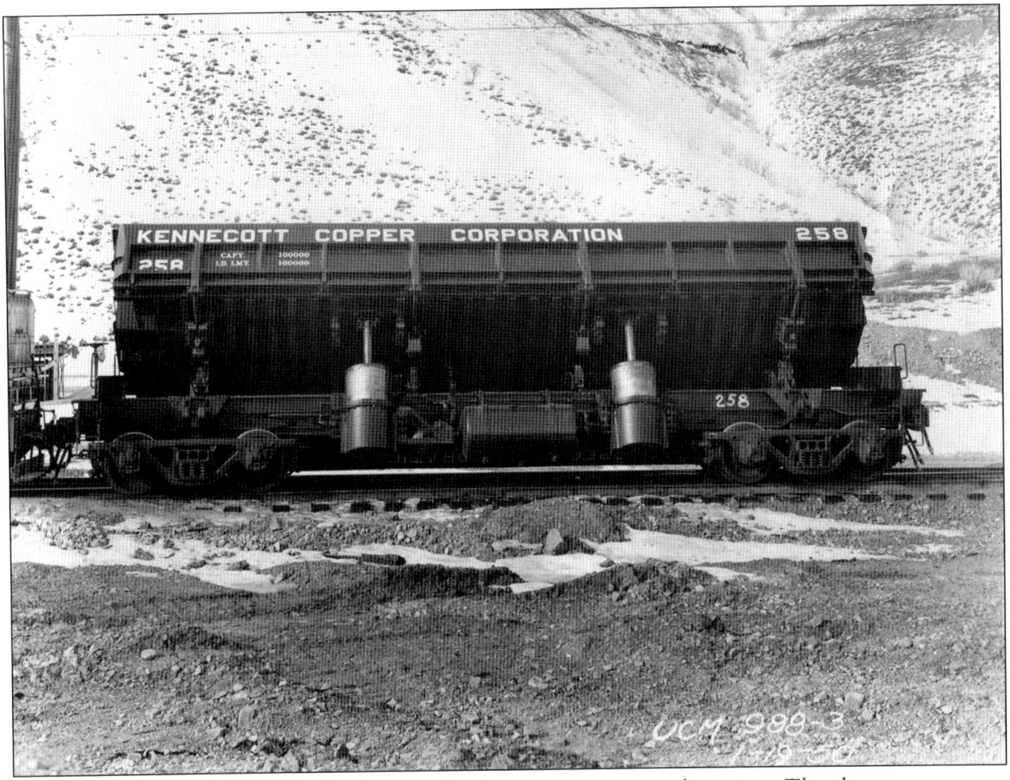

This 1950 image shows a new 40-yard side dump waste car at the mine. The dump waste cars were air-operated—note the large air cylinders. The first dump cars used at the mine were six-yard wooden and four-yard wooden narrow-gauge cars. It was not long before 12-yard all-steel waste cars were purchased. In 1961, the mine was working 33 waste shovels using 258 forty-yard waste cars.

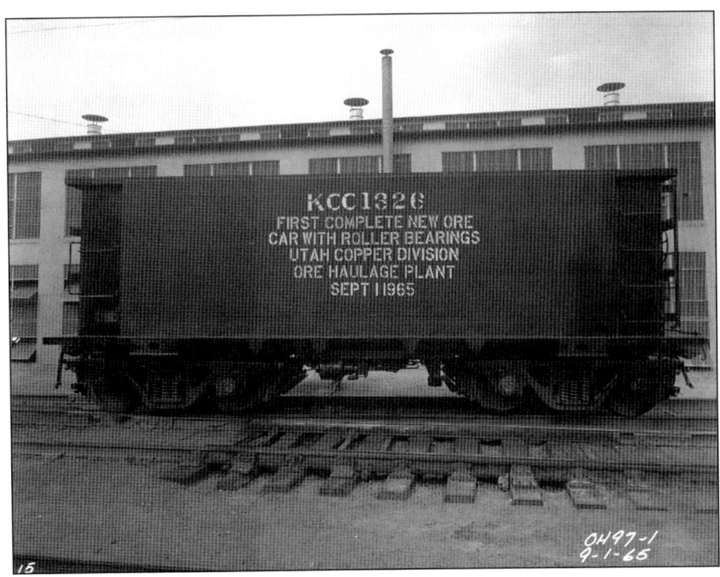

Pictured is a new Kennecott-constructed ore car on September 1, 1965. February 1963 was when Kennecott started constructing its own ore cars, using jigs and all-welded construction. In March 1972, Kennecott had 1,000 hundred-ton ore cars in service. Each ore car made on average 118 trips per month. Trains of 18–20 cars from the mine were gathered at Copperton yard into larger trains of 70–80 cars for the trip to the mills.

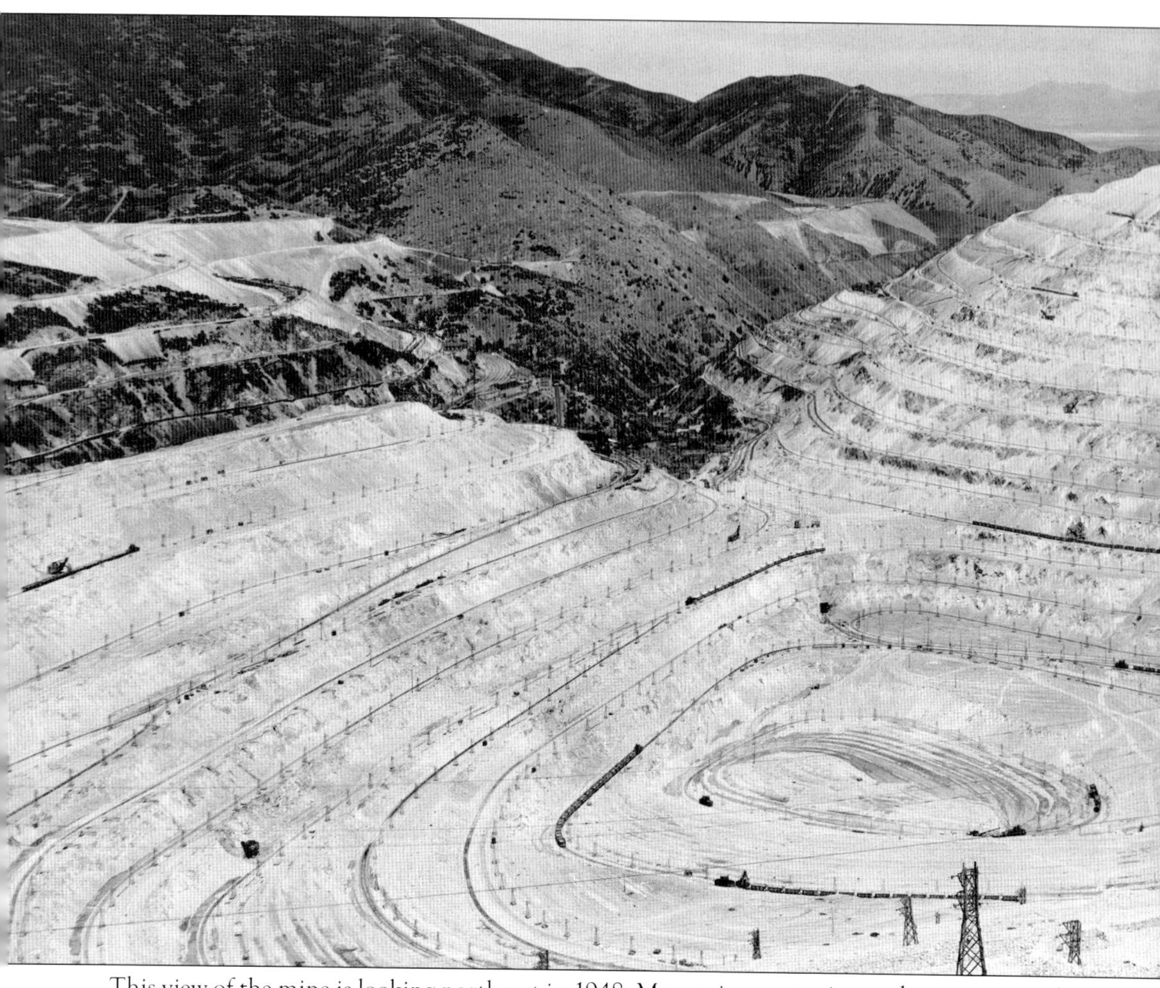

This view of the mine is looking northeast in 1948. More mine expansion and improvement have taken place. In 1939, a new vehicular tunnel was built to Copperfield, which enabled the mine to expand to the east. In April 1935, the mine had moved more material than when building the Panama Canal. In 1936, Utah Copper's name was changed to Kennecott Copper. In 1943, during World War II, copper was in great demand for defense purposes. Kennecott set these world records:

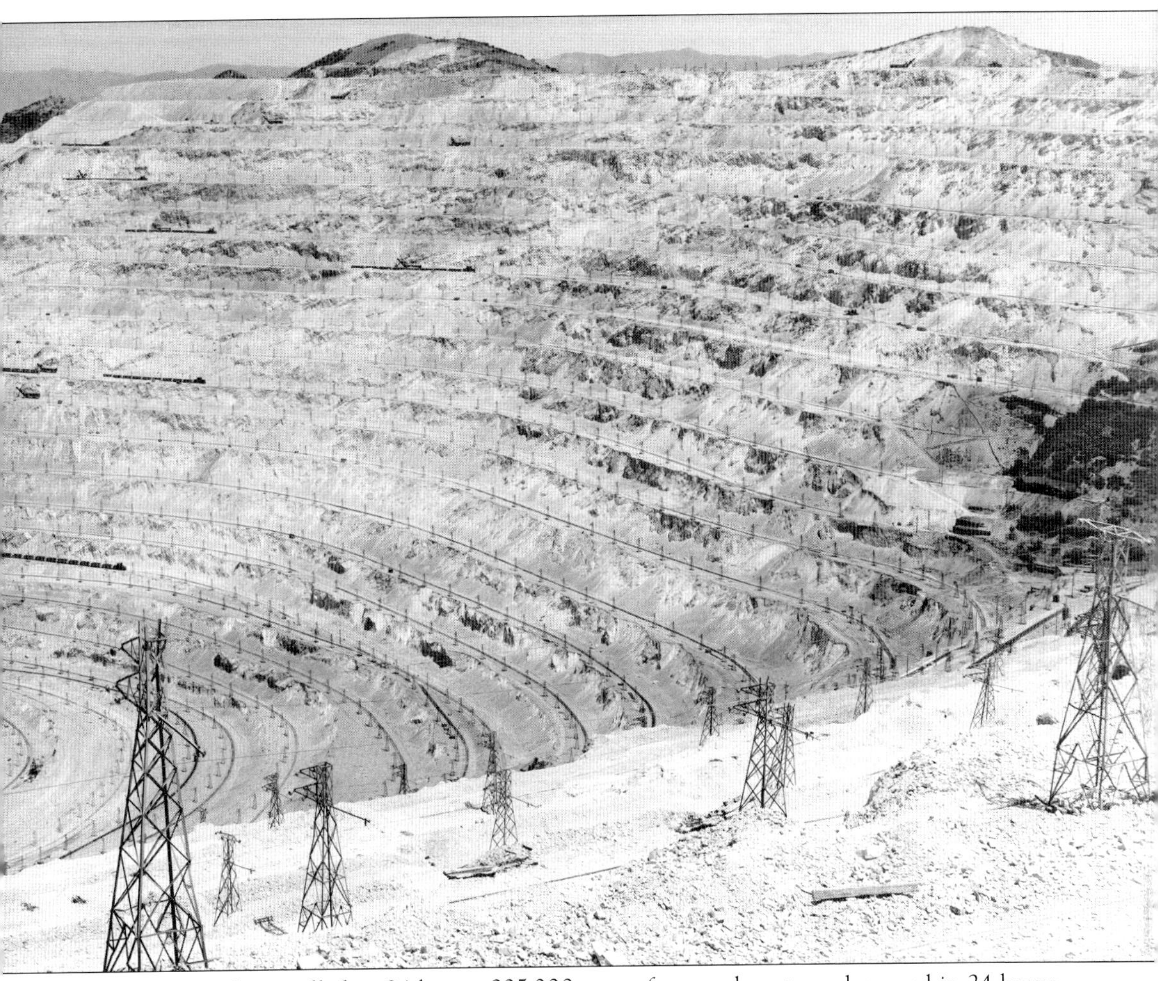

108,000 tons of ore milled in 24 hours, 235,000 tons of ore and waste rock moved in 24 hours, 3,074,000 tons of ore milled in one month, 35,375,000 tons of ore milled in one year, and 319,500 tons of copper produced in one year. In 1946, the first of three railroad tunnels was completed and named 6040, in relation to its elevation in the mine.

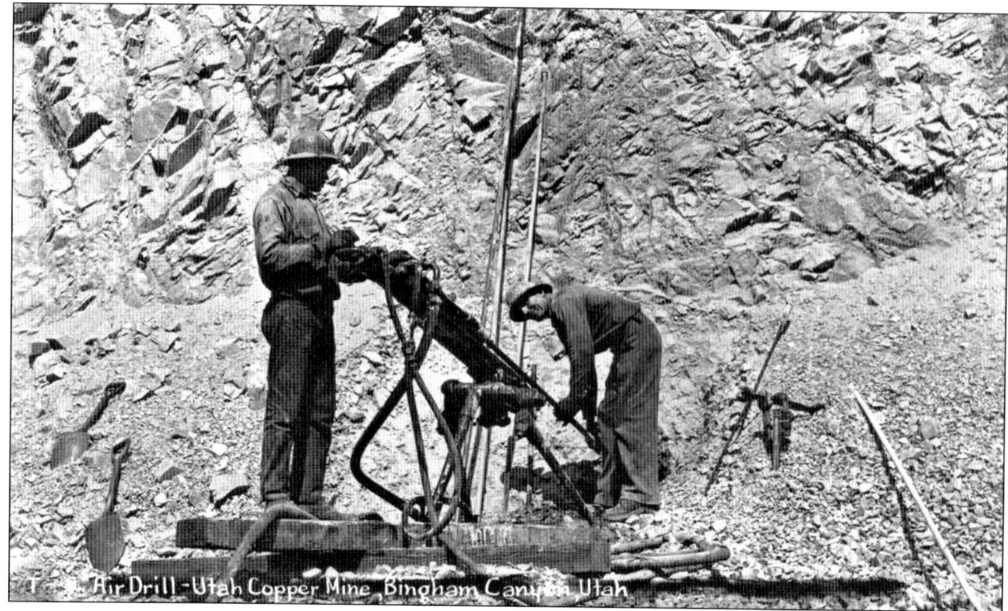

This image shows two men from the drill and blast department drilling holes at the base of the level. The pneumatic drill is set up on a tripod. This type of drilling is for spring hole blasting, which makes a large chamber hole. Then, a churn drill working on top of the level bores down to the chamber hole so a large amount of dynamite can be loaded into it.

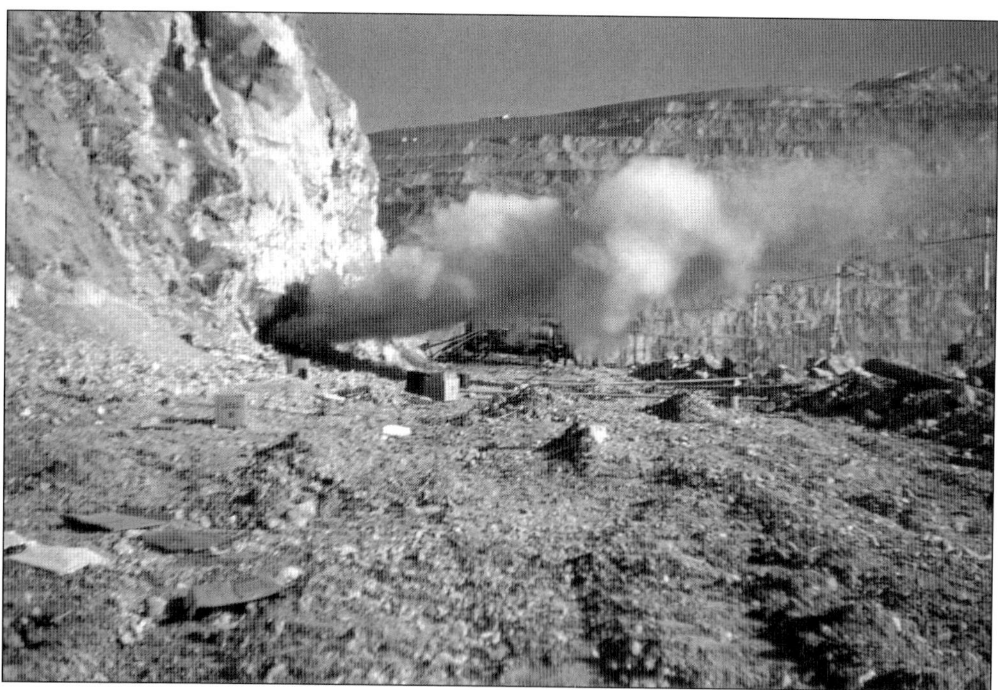

This image shows a spring hole blast. Once the spring hole is drilled, a two-man team loads the hole with dynamite. One man fills a pipe hole with one stick of dynamite at a time, while the other man pushes the sticks of dynamite deep into the pipe with a long wooden pole. The spring hole blast was a preliminary explosion for a larger blast.

Shown is a large blast hole drill in August 1972. Advancements in blasting techniques have changed over the years, improving greatly. No longer did men hang over the levels on ropes to push down loose rocks or drill spring holes at the bottom of the level in all kinds of weather. The drill and blast department now had large blast hole drills, like a Bucyrus-Erie 49-RII rotary drill, with a mast height of 97.6 feet.

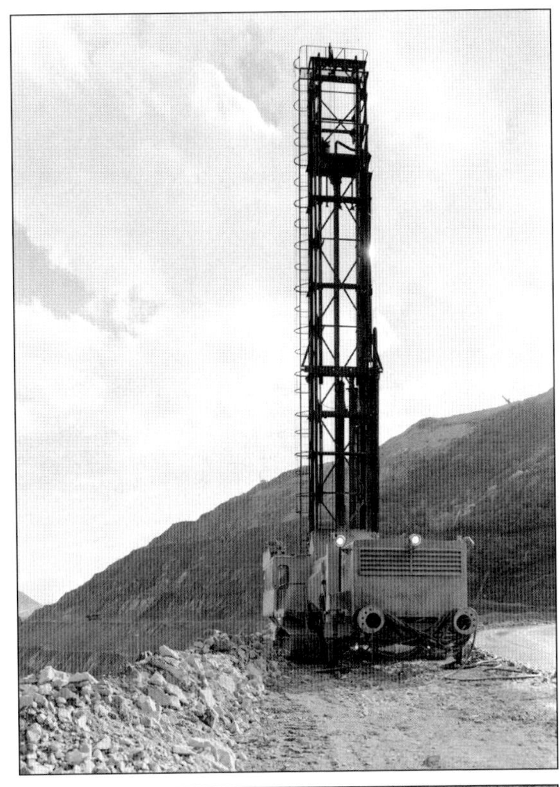

This photograph shows two men loading spring blast holes with a new blasting slurry instead of sticks of dynamite. Once the large rotary blast hole drills came to the mine, all the holes were loaded from on top of the level. In the 2020s, the blast at the mine took place one to four times a day by a highly trained staff. A detonation of 1,200 pounds of explosives in each 55-foot-deep blast hole would take place.

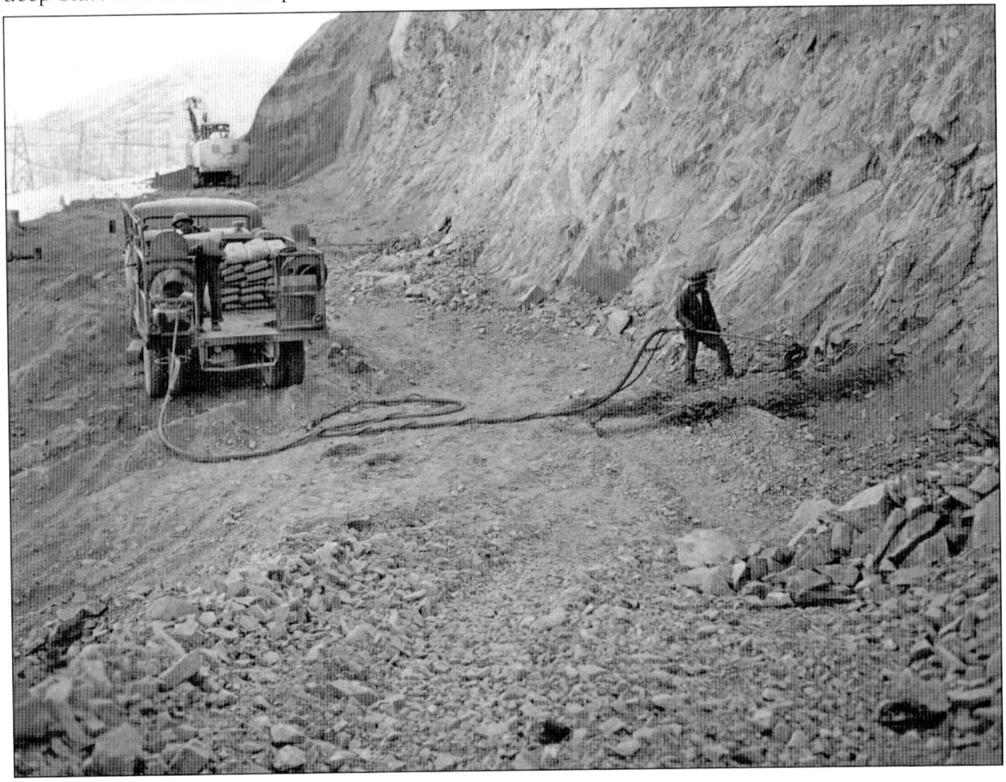

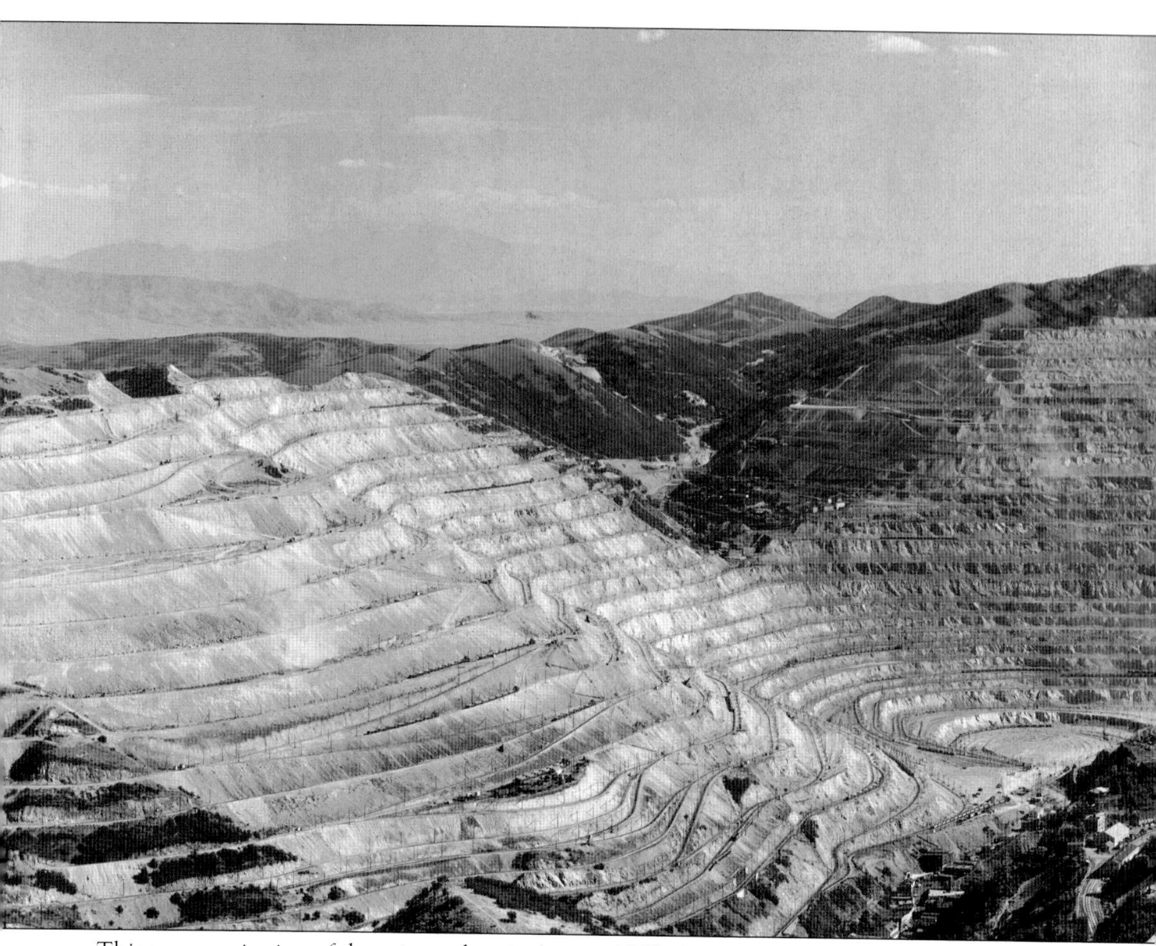

This panoramic view of the mine, taken in August 1950, shows it expanding in all directions. The towns around the mine were eventually gobbled up by that expansion. This view is looking south. The town of Copperfield, seen center left, was the first to go. The community of Highland Boy and Carr Fork Canyon, out of view to the right, were dug out. The town of Bingham, bottom center,

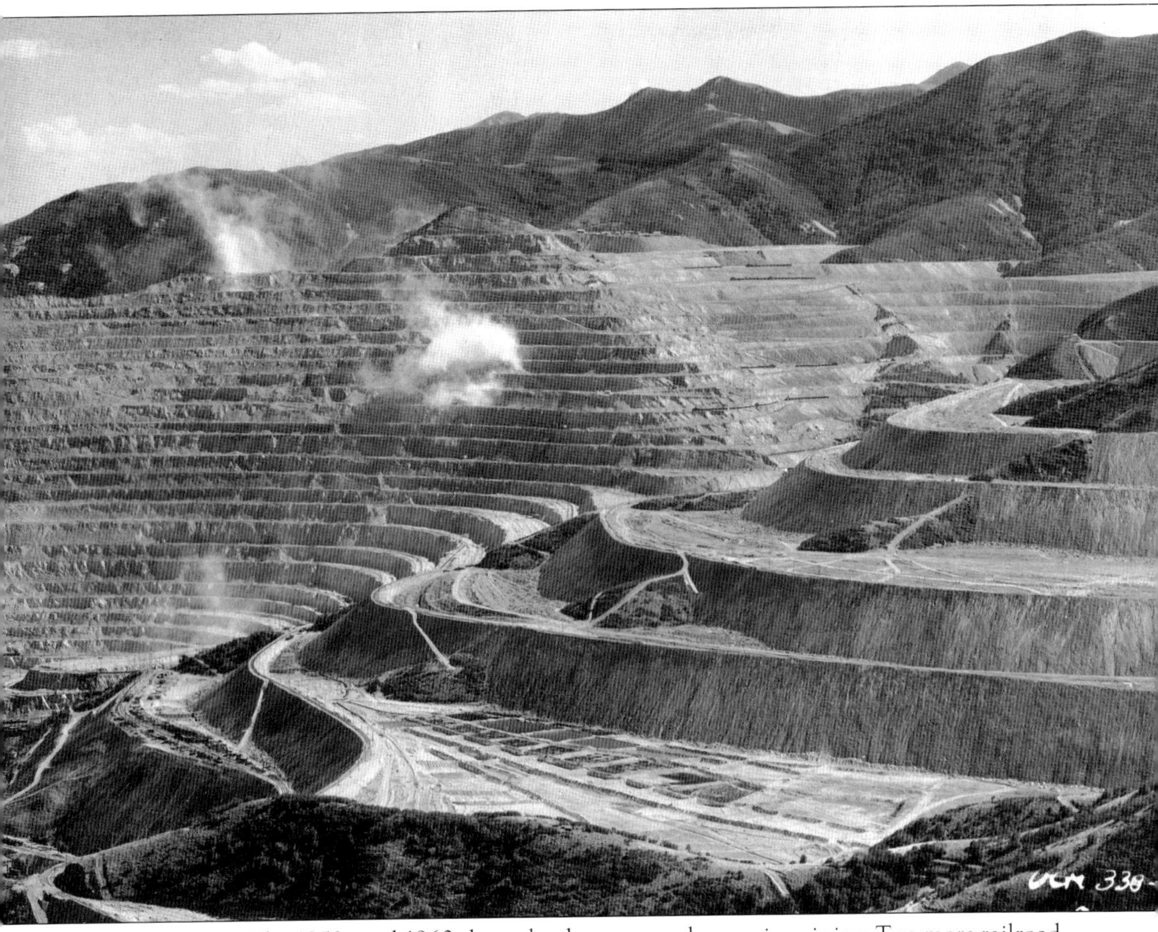

was the last to go. The 1950s and 1960s brought about more changes in mining. Two more railroad tunnels in the pit were dug, making three at 6,040, 5,840, and 5,490 elevations. The tunnels helped keep the trains moving downhill, lowering the cost per ton. The train shops moved down to Dry Fork Canyon, and the cross-canyon connection and a new central yard were built.

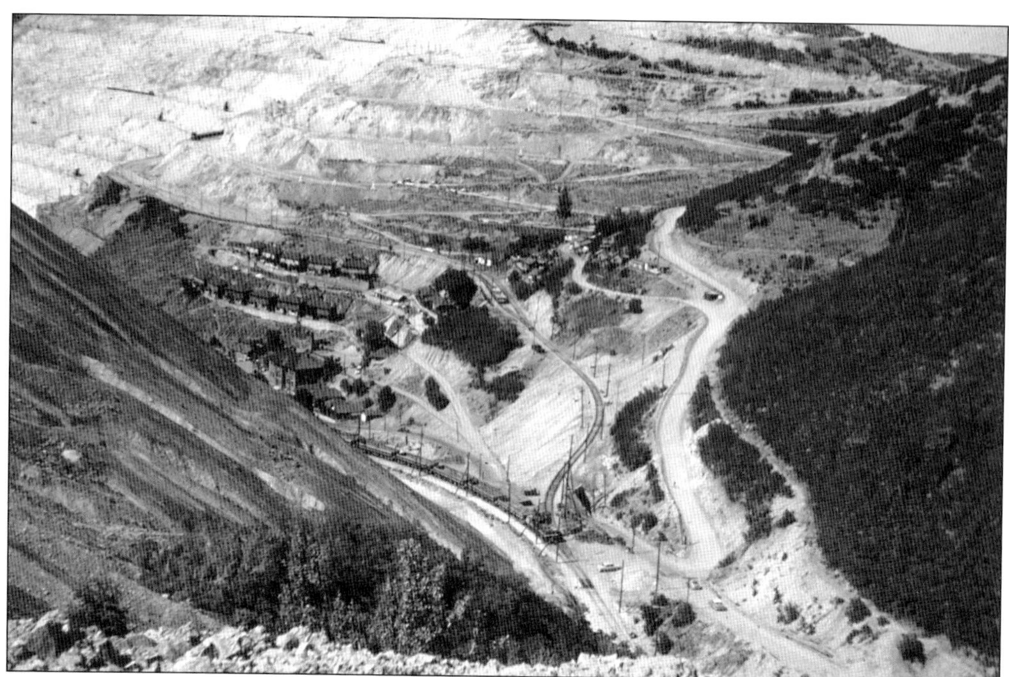

This image shows the small community of Copperfield in the 1950s. The encroaching mine is coming this way, and by 1958, Copperfield was demolished. This area was called "Upper Bingham," and in 1914, it was incorporated as Copperfield. Up Galena Gulch was the Old Jordan Mine, the first claim to be staked at Bingham. In 1899, mines in this vicinity were consolidated, becoming the US Mining Company mines.

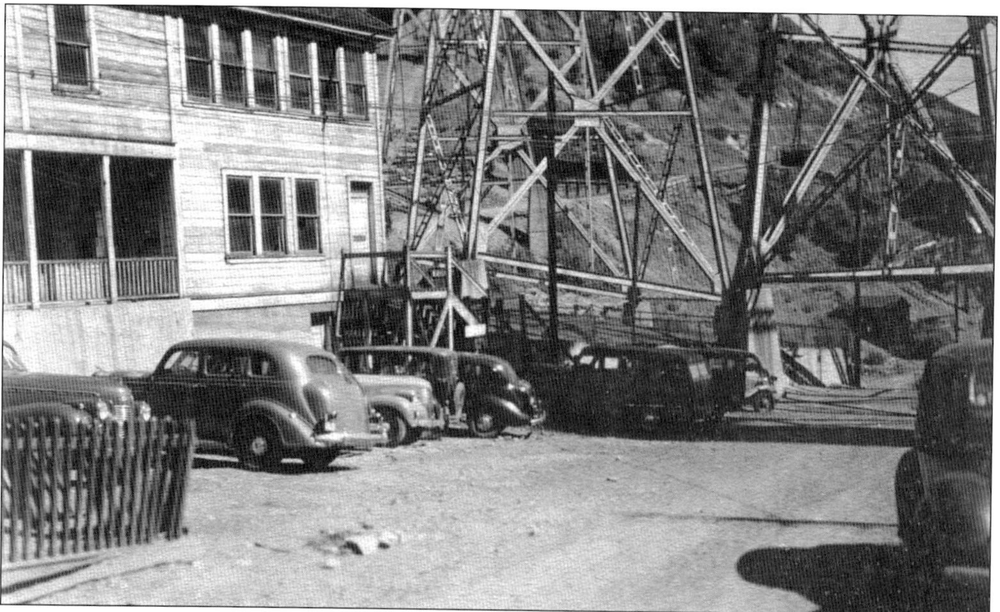

These cars and buildings are next to the L-line bridge in Highland Boy, the next community to be consumed by the mine. Residents received eviction notices on June 7, 1957. The seven large viaducts that once filled Carr Fork Canyon were removed. Two truck-to-train reload stations were built here because of the easy access to the rail lines that went down the canyon.

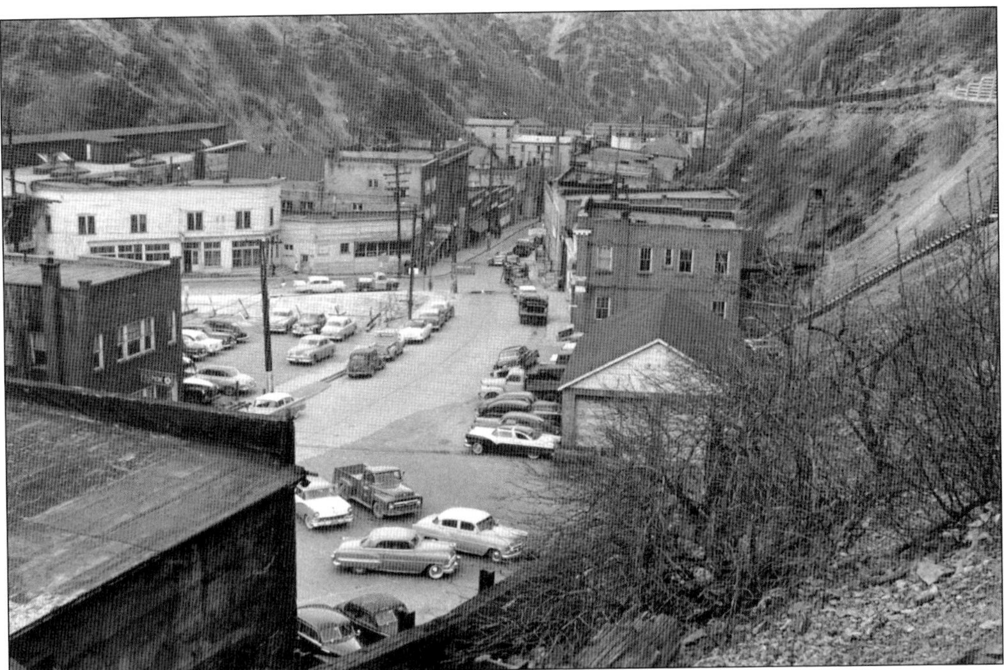

The town of Bingham, by the confluence where the road splits off to Highland Boy, is seen here in the 1950s. The Bingham Merc, once on the corner of Main Street and Carr Fork, was razed in January 1957. The cars are lined up, waiting their turn to enter the one-way Copperfield tunnel, seen at the bottom center. By 1971, the town of Bingham had disincorporated, with only a few people remaining.

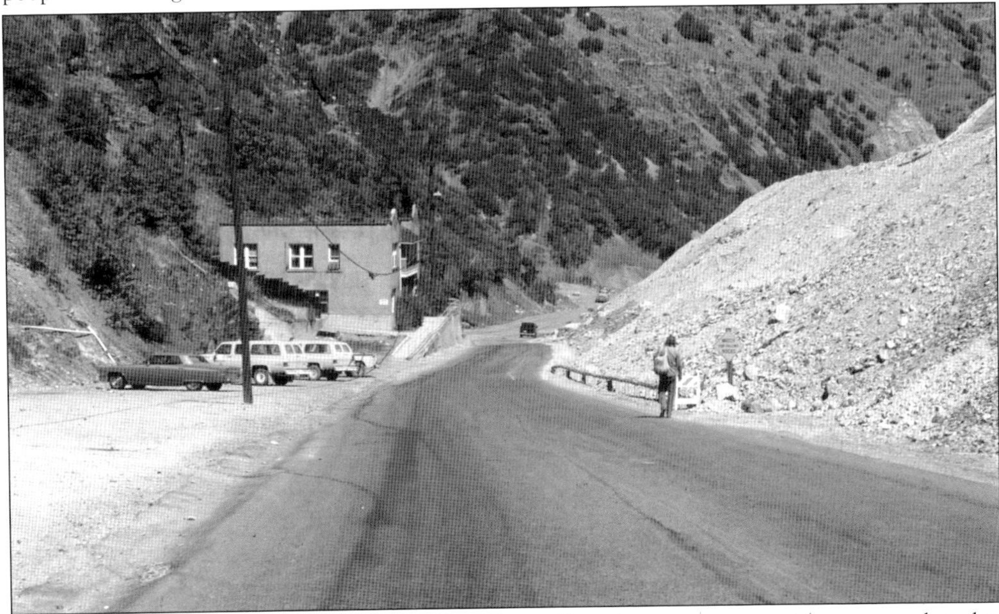

A lone man walks down the road where the town of Bingham once was, with waste rock to his right. By this time, there were only two buildings left: the clinic, as seen in the photograph, and the Gemmell Club, now used for mine offices. Both would be removed soon. Over time, the town of Bingham was dug away or filled with tons of overburden.

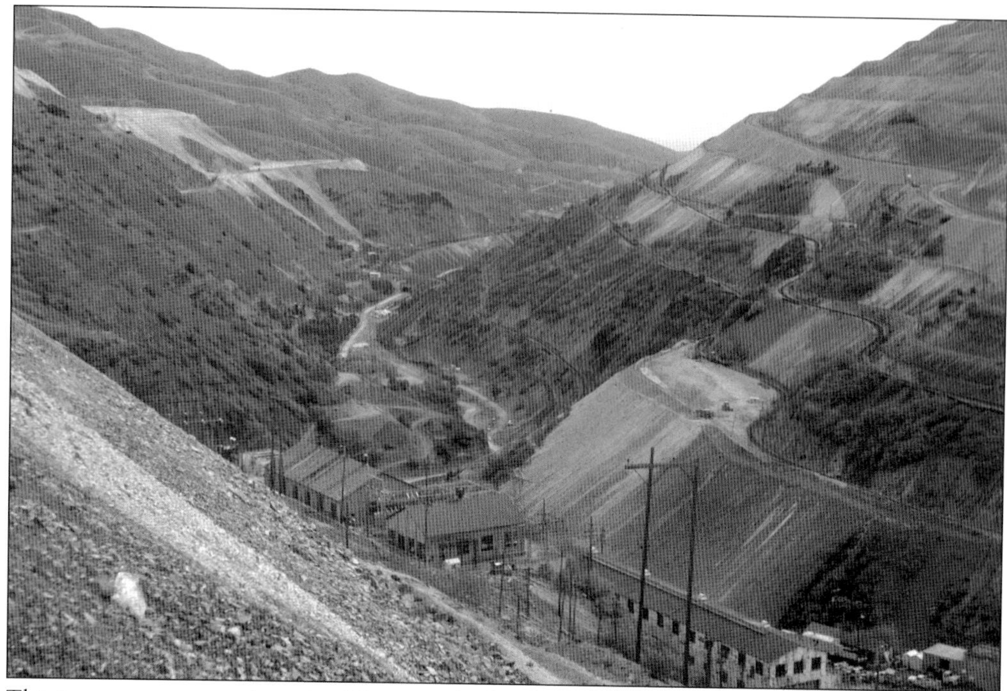

This image was captured north of the mine, looking down Bingham Canyon in 1975. The canyon is now void of people and buildings. The main canyon is being filled in with waste rock. This photograph was taken above the old B&G train yard, the shifter shed, and the oil house, seen bottom right. On the level above is the substation where AC electricity was converted to 750-volt DC current for the electric locomotives.

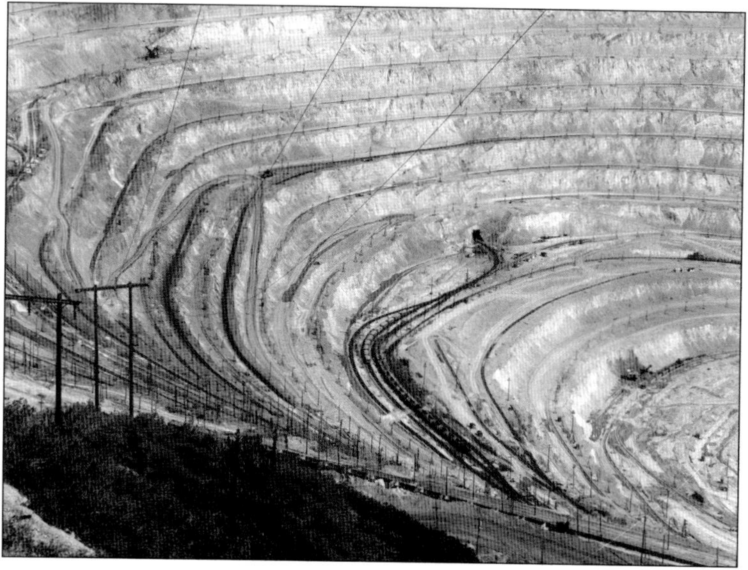

This image, taken in August 1972, shows the bottom of the pit. The 5490 train tunnel is a busy place with loaded ore cars being stacked up and empty ore cars coming into the mine, seen in the center. Trains are being used at the bottom of the mine; because of the steep grade out, only five cars could be pulled by the electric locomotives at one time.

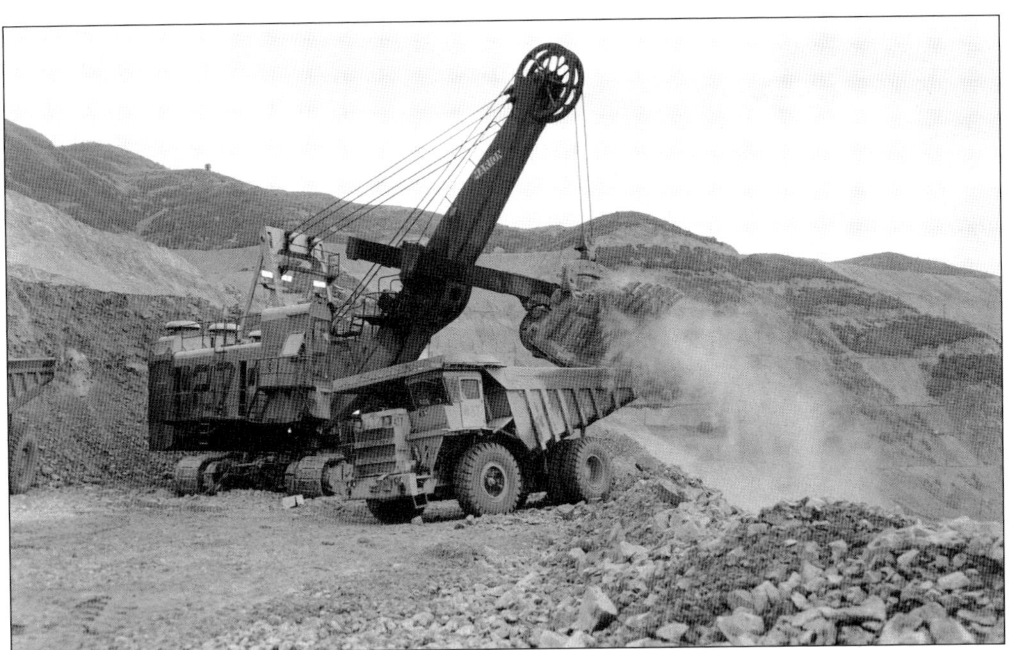

The No. 437 Haulpak truck is being loaded at the No. 27 electric shovel, working high on the west side of the mine by Carr Fork Canyon. An old aerial tram building is on the mountain in the background, seen center left. The 65-ton Haulpaks and KW Dart trucks started to be used at the bottom of the mine, as the trucks were small enough to make the trip down into the pit.

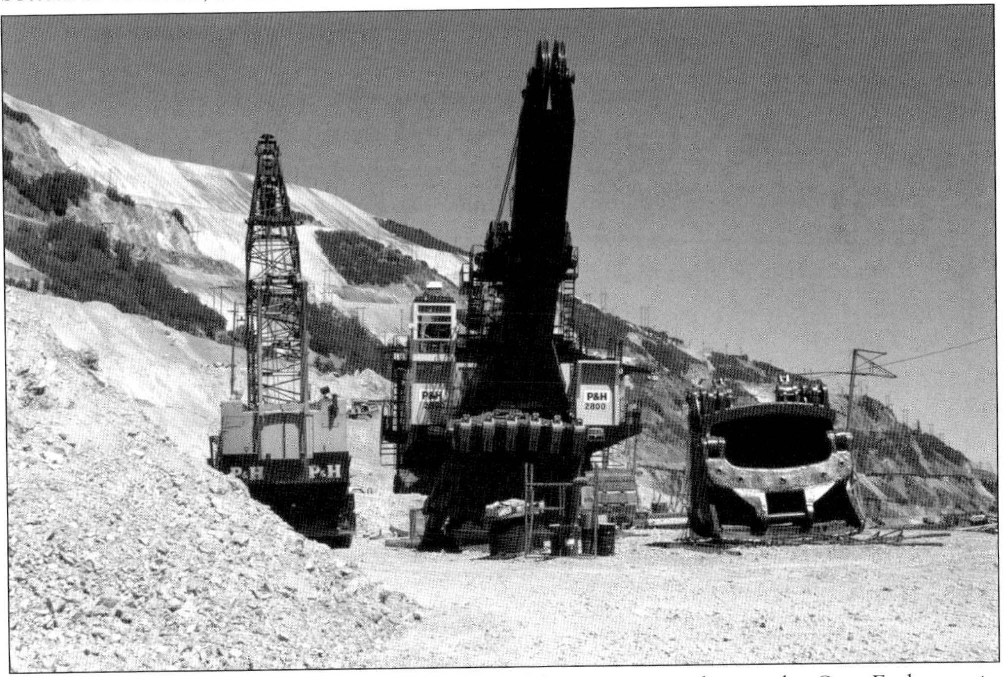

This image shows the No. 38 electric rope shovel being put together in the Carr Fork area in August 1975. The shovel parts came to the mine on rail flatcars. It is the first of four shovels this size coming to the mine. The shovel has a 25-yard dipper on it, and before it even enters service, a 27-yard dipper will replace it, as shown right. The new shovel replaced three six-yard shovels.

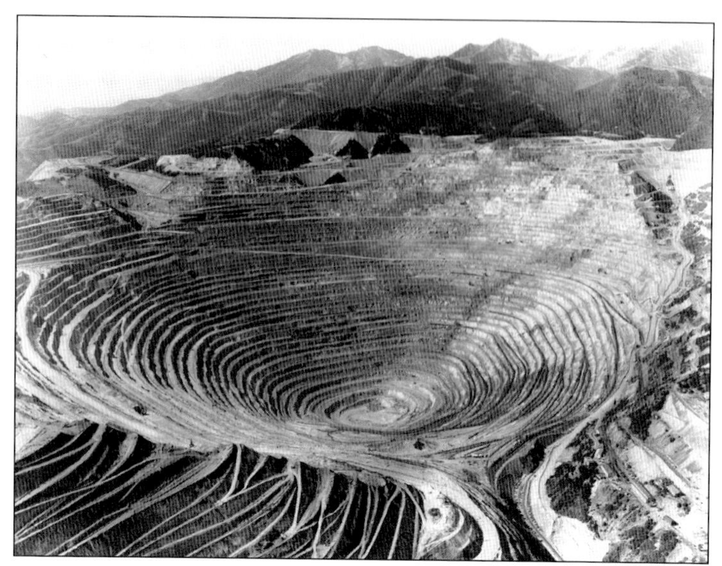

This image shows the mine on July 9, 1976. Truck haulage mines the top two-thirds of the mine, mostly removing overburden. Trains are still in use in the bottom one-third of the mine, moving ore. On September 21, 1972, the mining operation set a world record for material handled in a 24-hour period: 502,575 tons. The mine broke that world record in 1981, moving 638,649 tons. Bingham Canyon is bottom right.

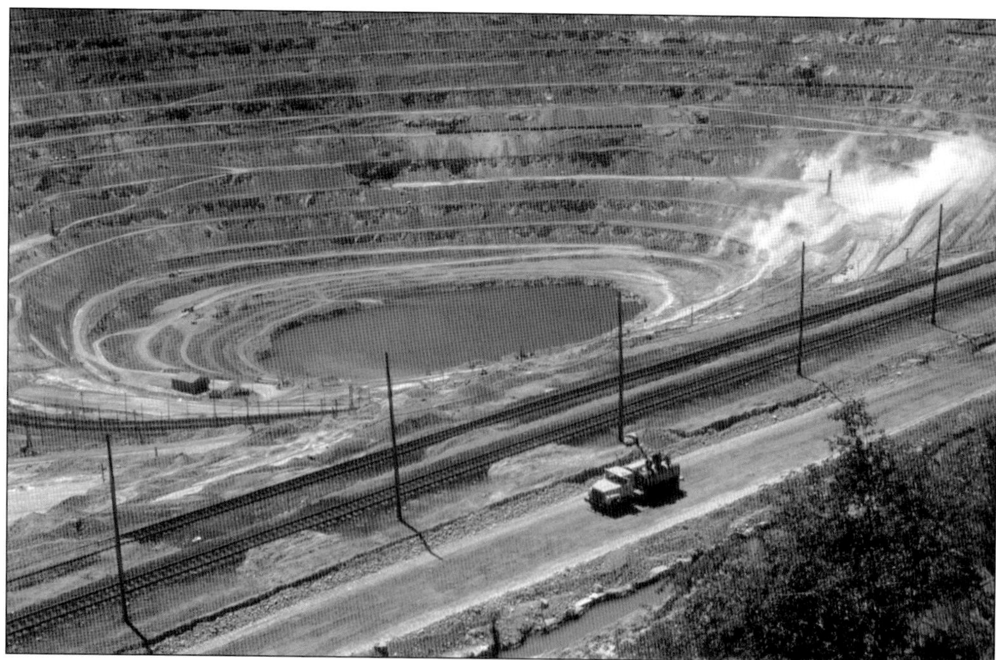

Kennecott's open-pit mine is seen here on July 3, 1983, as water fills the bottom of the mine. Water in the mine was always a problem and was constantly being pumped out. In 1983, the mine was converted to a shovel-to-truck operation. Trains moved the ore from three reloads. In 1984, there were 7,400 employees, but most would be laid off. With the new $400 million modernization, only 1,800 employees were needed.

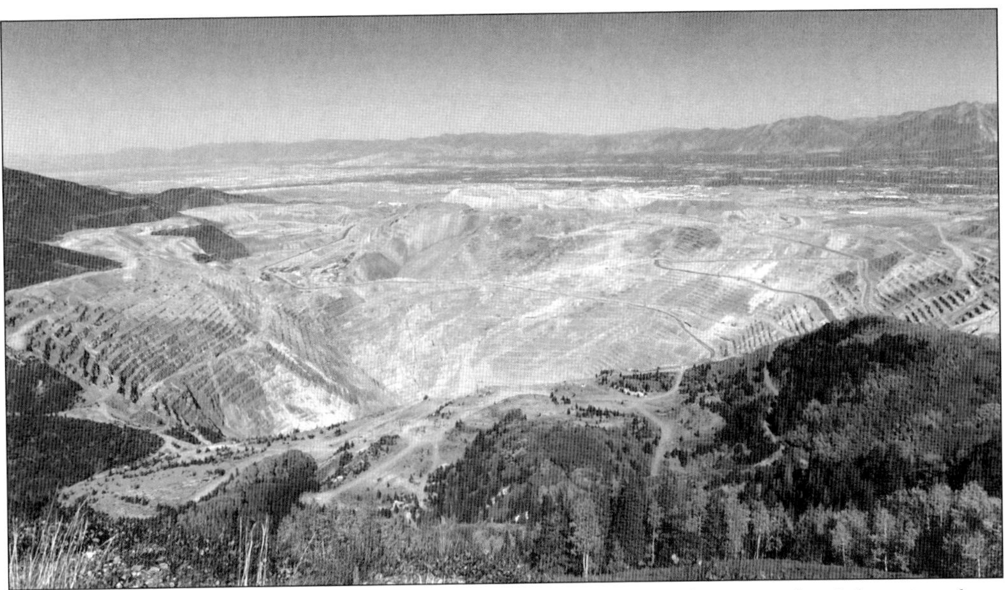

This image was taken in 2024 from the top of the mountain on the west side of the mine; from this vantage point, one is looking northeast. The mine spans 95,000 acres and is 2.75 miles wide and 3,900 feet deep. The smelting process produces copper, gold, silver, platinum, palladium, lead carbonate, and selenium. Molybdenum and tellurium are also produced. Research is being done to enhance the reclamation at the mine. (Author's collection.)

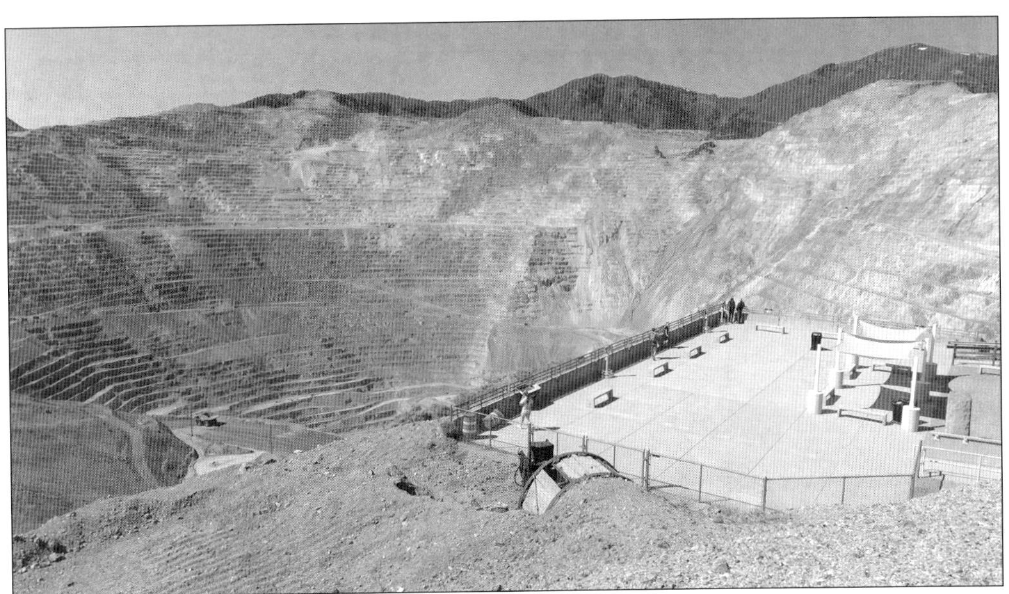

The Bingham Canyon Mine is seen here during a Rio Tinto Kennecott visitor experience on June 3, 2025. Rio Tinto Kennecott has invested billions of dollars into the cornerstone project and the south wall pushback. This south side of the mine will be dug back 1,000 feet, which will deepen the mine another 300 feet and keep it productive. Underground mining is going on and will move to the North Rim Skarn ore body. (Author's collection.)